# STRATEGIES FOR SUSTAINABLE MOBILITIES

# Strategies for Sustainable Mobilities

## Opportunities and Challenges

*Edited by*

REGINE GERIKE
*University of Natural Resources and Life Sciences, Vienna, Austria*

FRIEDERIKE HÜLSMANN
*Öko-Institut e.V. and Technische Universität München, Germany*

KATRIN ROLLER
*Technische Universität München, Germany*

ASHGATE

Published by
Ashgate Publishing Limited
Wey Court East
Union Road
Farnham
Surrey, GU9 7PT
England

Ashgate Publishing Company
110 Cherry Street
Suite 3-1
Burlington, VT 05401-3818
USA

www.ashgate.com

**British Library Cataloguing in Publication Data**
A catalogue record for this book is available from the British Library

**The Library of Congress has cataloged the printed edition as follows:**
Strategies for sustainable mobilities : opportunities and challenges / [edited by] Regine Gerike, Friederike Hülsmann and Katrin Roller.
p. cm.—(Transport and society)
Includes bibliographical references and index.
ISBN 978-1-4094-5489-2 (hardback)—ISBN 978-1-4094-5490-8 (ebook)—ISBN 978-1-4724-0410-7 (epub)
1. Urban transportation—Environmental aspects. 2. Urban transportation—Social aspects. 3. Choice of transportation. 4. Transportation demand management. I. Gerike, Regine. II. Hülsmann, Friederike. III. Roller, Katrin. IV. Series: Transport and society.
HE305.S845 2013
388.4—dc23

2013011973

ISBN 9781409454892 (hbk)
ISBN 9781409454908 (ebk – PDF)
ISBN 9781472404107 (ebk – ePUB)

MIX
Paper from responsible sources
FSC www.fsc.org FSC® C013985

Printed in the United Kingdom by Henry Ling Limited, at the Dorset Press, Dorchester, DT1 1HD

# Contents

# List of Figures

# List of Tables

# Notes on Contributors

**Veronique Van Acker** holds a Ph.D. in geography and is a post-doctoral researcher at the Geography Department of Ghent University and an assistant professor at the Centre for Urban Studies of the University of Amsterdam. Her work focuses on the interaction between the built environment and travel behaviour, taking into account subjective influences such as lifestyles and attitudes.

**Kay W. Axhausen** is Professor of Transport Planning at ETH Zürich. His focus of work is the measurement and modelling of travel behaviour, with a special emphasis on agent-based simulations and GPS-based surveys of behaviour.

**Gamze Dane** holds a M.Sc. in geographical information systems and is an urban planner and current Ph.D. researcher at the Eindhoven University of Technology, the Netherlands. Her research interests include urban and transport planning, modelling of activity-travel behaviour and data collection techniques.

**Ben Derudder** is Professor of Human Geography at Ghent University and Associate Director of the Globalization and World Cities research network (GaWC). His research focuses on the analysis of transnational urban networks and the (persisting) importance of business travel in the space economy.

**Fred Dotter** is a senior project manager at Austrian Mobility Research in Graz. His research interests include all aspects of sustainable urban mobility with a focus on mobility management and networking for a more sustainable and environmentally friendly mobility.

**Katrin Dziekan** was a researcher at the Berlin Institute of Technology in the Department of Integrated Transport Planning and the project evaluation manager in CIVITAS MIMOSA. She now heads the section on environment and transport at the German Federal Environment Agency (UBA) in Dessau-Roßlau, Germany.

**Martin Fellendorf** is Chair of the Institute of Highway Engineering and Transport Planning at the University of Technology in Graz, Austria. His research areas are transport modelling and simulation.

**Regine Gerike** chairs the Institute for Transport Studies at the University of Natural Resources and Life Sciences (BOKU) in Vienna, Austria. Her research interests include all aspects of transport ecology and economy, including modelling of environmental effects, empirical research and holistic approaches to evaluating sustainable mobility.

**Margaret Grieco** holds Britain's first Professorship in the field of Transport and Society at Edinburgh Napier University. Educated at the University of Oxford, Margaret has published extensively in the field of transport and society. She has also undertaken research for various governmental and international development agencies on issues concerning transport and society.

**Stefan Hausberger** is Head of the Emissions Research Area at the Institute for Internal Combustion Engines and Thermodynamics, University of Technology in Graz, Austria. His research areas include measurement, simulation and reduction of vehicle-related emissions and energy consumption.

**Friederike Hülsmann** holds a M.Sc. in environmental engineering, a diploma in economics and was a researcher at Technische Universität München, Germany. She is now working at the Öko-Institut e.V. – Institute for Applied Ecology in Berlin. Her research interests include the modelling of travel behaviour and environmental effects, multimodal transport concepts and alternative vehicle technologies.

**Dave Horton** is a sociologist, writer and consultant based in Lancaster, UK. His research interests centre on everyday life, culture, sustainability and mobility, with his core expertise being cycling. He is a founding member of the Cycling and Society Research Group and co-editor of Cycling and Society (Ashgate 2007).

**Tim Jones** is a Senior Research Fellow at the Oxford Institute of Sustainable Development and the Department of Planning at Oxford Brookes University. His current research interests centre on travel behaviour, travel experience and particularly the impact of built environment design on walking and cycling.

**Elaheh Khademi** holds a M.Sc. in transportation engineering and planning and is an urban planner and Ph.D. candidate at the Eindhoven University of Technology, the Netherlands. Her research interests include transportation planning, modelling of activity-travel behaviour and studying the effects of pricing policies.

**Benjamin Kickhöfer** holds a diploma in industrial engineering and management from the Technische Universität Berlin and is currently working in the Transport Systems Planning and Transport Telematics research group. His main research interests comprise of policy analysis, emission modelling, discrete choice modelling, public acceptance and participation, and optimal public transport supply.

**Johanna Kopp** is a geographer and research scientist at ETH Zürich. Her research interests include all system relationships between transport infrastructure, politics, economics, innovative mobility concepts and travel behaviour.

**Karin Kraschl-Hirschmann** is a Scientific Assistant and Ph.D. candidate at the University of Technology in Graz, Austria. Her research focuses on modelling and analyses traffic flows and environmental effects.

**Giuseppe Liguori** graduated in law and specialized in European Affairs and European project management. Since 2009 he has worked as project officer at SRM, the Public Transport Authority in Bologna, and has been involved in the implementation of projects for sustainable transport and mobility as well as ICT development.

**Raphael Luz** is Scientific Assistant at the Institute for Internal Combustion Engines and Thermodynamics at the University of Technology in Graz, Austria, where he is working in the division on emissions. His field of activity focuses on emission modelling and software programming.

**Andrew Miles** is a reader in sociology at the ESRC Centre for Research on Socio-Cultural Change (CRESC), University of Manchester, UK. His interests include cultural participation, social and spatial mobilities, and the methodological and policy frames informing socio-cultural research.

**Niamh Moore** is a Research Fellow at the ESRC Centre for Research on Socio-Cultural Change (CRESC), University of Manchester, UK. She is an interdisciplinary feminist researcher with interests in movement, mobilities, sustainability, and transdisciplinary and qualitative methods.

**Stewart Muir** is a social anthropologist at the ESRC Centre for Research into Socio-cultural Change (CRESC), University of Manchester, UK. His recent research explores connections between utopias and imagined futures, mobilities, sustainability and everyday life.

**Caroline Mullen** works at the Institute for Transport Studies, University of Leeds, UK. She focuses on governance, transport policy and ethics of transport, especially on carbon governance and walking and cycling. Her research is multi-disciplinary and involves philosophy, policy analysis and social science.

**Petter Næss** is Dr. Ing., Architect and Professor in Planning in Urban Regions at Norwegian University of Life Sciences, Norway. Combining qualitative and quantitative research methods, Næss has carried out research for more than two decades on the influence of urban structures on travel behaviour. Other research interests include the driving forces of urban development, planning theory and the philosophy of science.

**Colin Pooley** is Emeritus Professor of Social and Historical Geography at the Lancaster Environment Centre, Lancaster University, UK. His recent research has focused on aspects of migration and mobility in the past and present, including the role of walking and cycling.

**Dora Ramazzotti** is an administrative expert. In 2002 she specialized in European projects and obtained a Master's degree in Euro-projecting at the International University Centre in Florence, Italy. She has been a General Affairs Manager at SRM, the Public Transport Authority in Bologna, since it was established in 2003, where she has managed several transport-related EU projects.

**Katrin Roller** is a sociologist and Ph.D. candidate at the Technische Universität München, Germany. Her research focuses on social science-based mobility research using qualitative and quantitative methods.

**Joachim Scheiner** holds a Ph.D. in geography. He is a senior researcher and lecturer at the Department of Transport Planning at the Technische Universität Dortmund, Germany. His research focuses on travel behaviour in the context of residential mobility, spatial development and societal change.

**Griet Scheldeman** is a social anthropologist working in the UK and Europe. Her research interests centre on perceptive and creative processes in the relationships between people and their environments. She is currently investigating creativity in Arctic scientists' field practices.

**Emanuele Strano** is a Ph.D. candidate at the Laboratory of Geographic Information Systems at the Ecole Polytechnique Fédérale de Lausanne, Switzerland. His research focuses on the interface between urban geography and advanced spatial analysis, with a particular focus on spatial complex networks.

**Miles Tight** is Professor of Transport, Energy and Environment at the School of Civil Engineering, University of Birmingham, UK. His research interests lie in the area of sustainable transport, with a focus on walking and cycling and on promoting long-term, large-scale changes in transport.

**Harry Timmermans** is a professor in spatial planning at the Eindhoven University of Technology, the Netherlands. His main field of research is the development of decision support systems for spatial planning.

**Peter van der Waerden** is a senior research associate at Eindhoven University of Technology, the Netherlands. His main field of research is the relationship between the built environment and individual travel behaviour.

**Frank Witlox** holds a Ph.D. in Urban Planning from the Eindhoven University of Technology, the Netherlands. Currently, he is Professor of Economic Geography at Ghent University. He is also a visiting professor at ITMMA (Institute of Transport and Maritime Management Antwerp) and an Associate Director of GaWC (Globalization and World Cities, Loughborough University). Since 2010 he has also been the Director of the Doctoral School of Natural Sciences (UGent).

**Dujuan Yang** holds a M.Sc. in management science and technology, and is an economist and Ph.D. candidate at the Eindhoven University of Technology, the Netherlands. Her research focuses on the modelling of travel behaviour and energy effects using quantitative methods.

# Preface

This collection of papers stems from the international scientific conference 'mobil.TUM 2011: Making Sustainable Mobilities – Interdisciplinary Perspectives' held in Munich, Germany, in April 2011.

The goal of this conference was to bring together researchers with different disciplinary and cultural backgrounds working in the field of sustainable mobility. A broad range of issues was covered in keynotes and paper sessions with three main focus areas: 1) conceptual discussions about the meanings of sustainable mobility in the different disciplines; 2) strategies and instruments; and 3) institutions, policies and governance. This book mainly addresses the second area. Our aim is to achieve a better understanding of mobility behaviour and mobility needs in different contexts to develop innovative strategies and to advance modelling approaches which evaluate these strategies.

The goal is not to develop an ideal package of strategies to achieve sustainable mobility, but rather to present innovations in the different disciplines and fields and to show how each of them can contribute to keeping all people mobile – today and in the future.

We would like to thank the contributors for the time and effort they spent preparing their chapters and we hope that our readers will find these insightful and inspiring.

We would also like to acknowledge the support of those that made the mobil.TUM 2011 conference such a successful event. In particular, we wish to acknowledge the financial support of the German Research Foundation (DFG) and the Hans-Böckler-Foundation. We would also like to thank our scientific committee for their support, the Oskar von Miller Forum for hosting the conference, the participants for their lively discussions during the conference and the local organizing team for making this conference possible. Special thanks go to Chelsea Tschoerner who has done the proofreading for all the contributions in this book.

# Introduction
# Conceptual Framework for Developing Future Mobility

Regine Gerike, Friederike Hülsmann and Katrin Roller

## Challenges and perspectives for future mobility

Today's societies are more mobile than ever before. We buy coffee from Nicaragua and clothes from China; we meet friends and relatives over long distances; we travel to other continents for education, work or recreation; companies collaborate around the globe. In this 'shrinking world' (Dicken 1998), distances have become 'short' thanks to constantly improving transport systems and technologies and the very dynamic development of information and communication technologies. Mobility promises freedom and prosperity and is an important pre-requisite for a high quality of life. To be mobile is a way of being in the world (Urry 2000).

This high mobility of people and of goods forms the basis of today's globalization processes but also causes problems and unintended side effects. The transport sector is a main driver of the societal challenges we face. Scarce fossil resources and climate change are two examples of these challenges. Conversely, the transport sector is highly affected by these developments and the societal changes needed to meet these challenges.

Dependence on fossil fuels and information makes transportation systems increasingly vulnerable to shocks in energy markets and systemic failure. The transport sector accounted for 32 per cent of final energy consumption in EU-27 in 2008 with road as the most important transport mode (81 per cent), followed by air (14.5 per cent), railway (2.5 per cent) and inland waterways (1.7 per cent) (European Commission 2011). The transport sector accounts for nearly one quarter of global energy-related greenhouse gas (GHG) emissions (IEA 2012). In EU-27, transport contributed 24 per cent to GHG emissions from all sectors in 2009 with transport GHG emissions 27 per cent above levels in 1990 (EEA 2011). This trend is projected to continue in spite of improvements in vehicle technology. According to Skinner et al. (2010), transport's GHG emissions in 2050 will be 74 per cent higher than in 1990 and around 25 per cent above 2010 levels if there is no additional policy intervention. This increase is mainly due to growth in demand for maritime transport, aviation and road freight, but other transport markets are also projected to grow. For example, international tourist arrivals grew by 4.6 per cent from 2010 to 2011 and reached 983 million worldwide. Preliminary estimates suggest that international tourist arrivals will reach'a total of 1.8 billion by 2030. These developments show the dynamic nature of this large and fast-growing sector (UNWTO 2012). Other transport-related environmental problems include air pollution, noise and land consumption (EEA 2011).

In addition to these environmental effects, transport is the source of some unintended negative economic, social and political risks and effects. Advances in transportation technologies have helped to sustain networked patterns of social and economic life. However, this increase in mobility and travel also contributes to disintegrative social processes. Far-flung social networks and the mobility burden of commuting and business travel undermine traditional social structures such as families, neighbourhoods and communities (Schneider and Collet 2010). New inequalities, social exclusions and risks arise in connection with unequal availability of and accessibility to transport. For example, low-income population groups often live close to streets with high traffic volumes, spend more time at home and are, therefore, exposed to higher air pollution and noise compared to high-income households (see Gaffron 2012 for Germany and the UK).

Grieco (2006) distinguishes between three main groups of factors influencing social exclusion: place-based factors including transport infrastructure and the built environment; social-based factors, which mainly describe the position within a socially stratified society; and personal factors such as individual skills, perceptions and attitudes. These individual, social and physical factors are tightly interwoven. Social norms are constantly changing and because of that dynamic process, mobility requirements and practices change to a *hypermobile* way of life (Urry 2000, 2007). The unequal distribution of 'mobility as capital' (Kaufmann et al. 2004) causes the systematic exclusion of groups with less mobility.

## The concept of sustainable mobility

None of these social, environmental and economic challenges can be dealt with in isolation. Various interdependencies exist with synergies and conflicts between the effects. Only integrated approaches are suitable for keeping people mobile both today and in the future. The question is: are the concepts of sustainable development and sustainable mobility suitable visions to guide us? These concepts call into question post-fossil 'mobilities' that are efficient, environmentally-friendly and socially inclusive. They are, however, qualitative, vague and normative visions. Such broad concepts are at risk of becoming arbitrary. The term 'sustainable mobility' is used in various contexts in political debates, as well as in scientific literature. Often it is used without a concrete definition of what is meant by this term. Low public acceptance can result from this ambiguity. Sustainable mobility has been criticized in the past as being just a catchword, which disguises the opposing interests of those involved without having any real substance.

Interestingly, this vagueness is also a quality, which provides sustainable mobility with an advantage over other concepts. It allows for integrated strategies, which meet the societal challenges at hand. It allows diverse stakeholders to commit to the goal of sustainable mobility. It allows for consensus. This is consensus on an aggregated and abstract level but this consensus can also help us achieve a transport system that ensures mobility for current and future generations.

In fact, this consensus is quite impressive; many stakeholders from all parts of society throughout the world have committed to the goals of sustainable development and sustainable mobility. At the Conference on Environment and Development in Rio de Janeiro in 1992, 178 states agreed to this goal (United Nations 1992).

However, 20 years after Rio+20 we must face the sobering thought that little of the momentum resulting from Rio is still with us today and that the goal of sustainable development is as out of reach as it ever was. Thus, it is clear that an institutional framework for global sustainable development must be established (Earth Summit 2012) because universal agreement to goals and strategies of sustainability still remains a challenge.

In 1997, the EU included the goal of sustainable development in the Treaty of Amsterdam as an overarching objective of EU policies. In 2006, the sustainable development strategy was adopted (European Commission 2009). Sustainability is considered a guideline for all fields of politics in many countries.

Hence, from these political and institutional considerations, the concept of sustainable mobility seems to be suitable for meeting societal challenges and for guiding strategies which encourage our societies to ensure a high quality of life for everyone both now and in the future.

This fact holds true despite the difficulties in achieving this highly consensual vision. We see two main problems that are hindering substantial, target-oriented progress towards achieving sustainable mobility. First, the concept is often described too vaguely. Second, the concept is ambitious especially due to its global dimension; global institutions need to be involved in order to push and coordinate the various activities.

In the following, we address the first of these two points: what do the concepts of sustainable development and sustainable mobility really mean? What is behind these vague visions?

The concept of sustainable development originally derives from forestry. Hans Carl von Carlowitz (1713) described forest management, which aims for long-term success in his book, *Sylvicultura Oeconomica*. Basically, no more wood should be taken from the forest than can grow in it again. The starting point for the present discussion on sustainability is mainly the work done by the World Commission on Environment and Development (WCED) and the 'Brundtland-Report' (WCED 1987), which uses the following definition:

> Sustainable development is development that meets the needs of the present without compromising the ability of future generations to meet their own needs. It contains within it two key concepts: the concept of 'needs', in particular the essential needs of the world's poor, to which overriding priority should be

> given; and the idea of limitations imposed by the state of technology and social organization on the environment's ability to meet present and future needs (WCED 1987).

The anthropocentric character of this definition is obvious: human beings are the focus of the definition. It follows the principle of intra- and intergenerational equity and claims to meet the needs of all people, including future generations. The commission remains vague about what is meant by 'essential needs'. The definition stresses the process character of the concept of sustainable development and excludes outcome-oriented policies. The essential needs of humans are subject to change for all people, today and in the future. Conflicting aims between different regions or social groups within a generation and between generations are likely to occur with this challenging approach.

Despite initial attempts to define the concept of sustainable development in concrete terms using the Brundtland definition the concept still remains vague. The challenge is how to make this vague vision manageable for transport policy.

There is a lively scientific debate on sustainable transport and mobility (e.g. Banister 2007a, 2007b, Gerike 2007, Gudmundsson 2004, Litman 2006). Various official and scientific indicator systems for sustainable transport and mobility have been developed (Ahvenharju 2004, Eurostat 2007, Litman 2008, OECD 2008, Wiederkehr et al. 2004). The COST 356 initiative 'EST – Towards the definition of a measurable environmentally sustainable transport' focused specifically on developing indicators for the environmental effects of transportation (Joumard and Gudmundsson 2010). The links between research in transport and mobility and research in sustainability sciences are only very loose (see e.g. Rauschmayer et al. 2011).

Building on this discussion, we will develop our view on the concept of sustainable mobility in this book based on the Brundtland definition. Sustainable mobility is an anthropocentric concept. Current and future generations, and their needs, are the focus of attention. Taking this perspective, sustainable mobility becomes an inherently open and process-oriented concept.

Human needs are complex and dynamic with various interdependencies and constantly changing weightings of specific needs. Therefore, the development of needs-oriented mobility will never be a linear process. Mistakes are a necessary part of the learning process. Doing, being and interacting allow us to realize and fulfil different needs in different ways (Rauschmayer et al. 2011). This autonomous and culturally embedded search for quality of life is an important part of human development. Different means of mobility are used to fulfil various needs – and the degree of fulfilment is unknown beforehand – even more so for future generations. Designing sustainable mobility is an ongoing task that can guide policy making today and in the future and that leads to a variety of sustainable mobility practices and systems.

Limitations are necessary to constrain this learning process and to avoid the arbitrary use of the concept. In transportation, limitations set for environmental effects are often exceeded. There is broad consensus that the current amount of natural resources used and the amount of greenhouse gases, noise and air pollutants emitted, cannot be regarded as sustainable. Limitations can also be formulated for social and economic dimensions. Here, the task is to ensure basic accessibility for all transport users and to develop transport systems that are affordable for current and future generations. The premise for social equality from an intra- and intergenerational perspective should be that everybody has access to activities, to mobility and to destinations in order to participate in social, economic and cultural life. This does not mean the equal distribution of goods but rather the equal distribution of opportunities for personal development.

Overall, we view sustainable mobility as an ongoing process that relates to basic human needs in an open and process-oriented way. The overall aim of sustainable mobility should be to give everybody the opportunity of being mobile in order to fulfil their needs while respecting the limitations of these transport and mobility practices.

## Outline of this book

The goal of sustainable mobility, as described above, is an ambitious one. It takes a long-term and process-oriented perspective. It can only be dealt with in a social discourse with various stakeholders including the transport users involved. The goal of this book is to support sustainable mobility practices from the researchers' perspective and is divided into three main parts as follows:

- *Activity patterns in different contexts:* Knowledge of mobility behaviour and its drivers are necessary to cope with societal challenges and all their interdependencies, goal conflicts and synergies. We need to understand the people and their (mobility) needs before being able to successfully develop and implement needs-oriented strategies. The first part of the book contains chapters, which analyse various aspects of mobility behaviour in different contexts.
- *Models for sustainability:* Decisions for transport often have long-lasting economic, social and environmental consequences. Transport models help to forecast these consequences and thus, contribute to transparent and efficient policy-making. Substantial progress has been made in transport modelling in the last few years. More complex mobility patterns can be modelled for larger areas and there is a dynamic community advancing this research. Chapters in the second part of the book present advancements in this field.
- *Innovative strategies:* The third part of the book presents innovative strategies and methods of evaluating these strategies. Many stakeholders all over the world are actively working on improving the sustainability of our transport systems. We describe experiments that have been done in different countries: who has done what and how and with which results?

The following section explains the book's three parts in more detail.

*Activity patterns in different contexts*

Strategies for sustainable mobilities can only be developed when people's needs in different contexts are known. Contributions in this part of the book describe and explain mobility behaviour: which social, socio-demographic, spatial and transport-system-related factors drive mobility behaviour? Understanding these drivers provides the framework for developing innovative strategies that are addressed in the third part of the book.

Jones et al., Næss and van Acker et al. analyse interactions between the built environment and mobility behaviour, that is, destination choice for activities, modal choice, route choice and activity participation. Jones et al. draw on quantitative and qualitative data generated from the large-scale UK study *Understanding Walking and Cycling* to investigate decision making in relation to walking and cycling in English cities. The authors conclude that dense networks for walking and cycling are a vital pre-requisite for achieving high modal shares of walking and cycling but that strategies for improving the quantity and quality of these networks need to be embedded in a multi-faceted approach so as to create safe and pleasant physical environments.

Næss presents insightful qualitative analyses for two case studies from Copenhagen (Denmark) and Hangzhou (China). Even though the cultural and economic backgrounds in these two cities are obviously very different, Næss identifies similarities in mobility patterns and underlying rationales. Interviewees ranked freedom of choice and the quality of destinations consistently high, so that mobility is influenced more by the distances between home and city centres with concentrated facilities than by the distance to the closest single facility within a category.

Van Acker et al. analyse objective and subjective spatial influences of mode choices for leisure activities. These analyses were based on a unique database including an Internet survey of lifestyles and leisure mobility in Flanders (Belgium), which also questioned the respondents' perceptions of their current residential neighbourhood and spatial information from other land use databases. With this approach, perceptions of residential neighbourhoods can be compared with objective spatial characteristics and the travel consequences of the (mis-)matched spatial perceptions can be evaluated.

Scheiner presents an analysis of time and work sharing in German households from 1994 to 2008. He addresses equity issues, which is an important and often underexposed aspect of sustainability. The author focuses on mobility in terms of travel time plus out-of-home activity time studied from a gender perspective. Traditional time-use patterns of couples decreased in the period studied in line with increasing employment among mothers. However, this development has not resulted in greater gender equality in terms of who does the housework.

*Models for sustainability*

Developing and improving models is traditionally a core field of transport research. Models for sustainability should be able to quantify not only changes in travel demand for different scenarios but also economic, social and environmental effects. Knowledge from different disciplines is needed to understand the various effects and their complex interdependencies. The contributions in the second part of the book present advancements in this type of integrated modelling ranging from microscopic to macroscopic approaches.

Hülsmann et al. and van der Waerden et al. use agent-based models to investigate the effects of pricing measures and changes in infrastructure. Hülsmann et al. compute the complete cause and effect chain from trip generation to determining air pollution concentration. They analyse changes in air pollution concentration for different pricing scenarios using a modelling tool developed for the city of Munich in Germany. Van der Waerden et al. analyse the effects of parking measures on shopping trips with the help of a parking analysis model called Pamela.

Kraschl-Hirschmann et al. develop a method of calculating energy consumption as an input for routing in navigation systems. This method is applicable to all types of cars including electric cars.

Khademi et al. describe the design of choice experiments, which study accumulated pricing schemes subject to budget constraints. This experimental approach studies travellers' responses to different combinations of transport charges, which is increasingly the case in real life, but requires complex models and experiment designs.

All four of these contributions enhance the tools and approaches used in modelling in order to better simulate travel behaviour and environmental effects.

Miles et al. identify the deficiencies in transport models, which are purely scientific by looking at the simplifications, ambiguities and normativity that are inherent in these approaches. They present an innovative approach to effective modelling, which includes mobility biographies. This critique of current methods of modelling together with a promising alternative is a valuable contribution to the discussion of evaluation practices in transport. However, it is more likely that these different approaches will complement rather than replace existing models.

*Innovative strategies*

Understanding people's (mobility) needs and mobility behaviour provides the basis for designing and implementing strategies for sustainable mobility. In this part of the book, we look beyond current sustainability strategies and present innovative ideas and pilot projects, which incorporate fresh perspectives on sustainable mobility and respond to new developments and technologies in the transport sector. The aim is to describe selective innovative strategies including institutional settings; critical factors of success or failure; details of implementation; and effects of the measures in terms of changes in travel demand as well as economic, social and environmental impacts.

The development and implementation of these strategies involves stakeholders from politics, administration and the civil society and thus goes beyond ideas that are purely scientific. These transdisciplinary working processes are a vital part of sustainable development that is open and process-oriented as described earlier. They are incorporated into this part of the book, for example, in Dziekan et al. and Dotter et al. The goal is to bridge 'the gap between problem solving and scientific innovation' (Lang et al. 2011) and to take the insights gained from research together with innovative ideas generated from participative and inclusive processes and put them into practice.

Dziekan et al. introduce a mobility credit system to identify effective incentives to encourage people to use more sustainable transport modes. Pilot tests of this system are described and the intended effect on travel behaviour and greenhouse gas emissions is evaluated.

Sharing instead of owning mobility tools is a promising approach to reducing car ownership and usage and for increasing the number of multi-modal transport users. Kopp et al. give an overview of the development of CarSharing systems with a focus on Europe and Germany. They then describe the new free-floating CarSharing system called DriveNow and present initial figures regarding users and usage. Kopp et al. show that DriveNow cars are mainly used for short distances and in parallel to other modes of transport. This innovative service is still too young to evaluate its effects on mode choice, destination choice and the environment.

Grieco focuses on issues of social exclusion and social equity in the case of crisis events. She discusses problems and insights gained from various case studies. Given the high number of disasters all over the world in the last few years and the resulting social, financial and ecological damage, this paper provides valuable insight into preventive spatial and transport planning, and catastrophe management that can be used to benefit different parts of the world.

Mobility management is widely accepted as a concept that promotes sustainable transport (Hyllenius et al. 2009). The focus is on 'soft' measures such as information. These measures should enhance the effectiveness of 'hard' measures such as changes in transport supply with both types of measures ideally being applied in package approaches. Hardly anybody doubts the value of mobility management but its evaluation is challenging because of complex cause and effect chains and potentially large spatial and temporal offsets between the implementation of measures and the appearance of their effects. Dotter describes and illustrates two innovative approaches to planning, monitoring and evaluating mobility management projects: MaxSumo and MaxEva. This transdisciplinary approach explicitly goes beyond purely scientific discussions of problem solving. It includes local stakeholders and considers local conditions when developing, implementing and evaluating strategies for mobility management.

## Summary

The contributions in this book are snapshots of different aspects of sustainable mobility. They cover a wide range of topics, disciplines and methodologies ranging from fundamental to transdisciplinary research. However, by no means does this book cover every area that is relevant to the concept of sustainable mobility. The papers deal with environmental, economic and social aspects; and they discuss all types of strategies including education, engineering, encouragement and enforcement. The level of analysis ranges from general overviews (e.g. Kopp) to very specific analyses (e.g. Kraschl-Hirschmann et al. and Khademi et al). All of these analyses, models and pilot studies are important and should form the basis for integrated solutions that help to reduce the unsustainability of our transport systems.

## References

Ahvenharju, S., Könnölä, T., van Grol, R., Walker, W., Röhling, W., de Tommasi, R., Arend, M., Bickel, P., de Ceuster, G. 2004. SUMMA (Sustainable Mobility, Policy Measures and Assessment). Operationalising Sustainable Transport and Mobility: The System Diagram and Indicators, Deliverable 3, http://www.tmleuven.be/project/summa/home.htm [accessed: 05 December 2012].

Banister, D. 2007a. *Unsustainable Transport. City Transport in the New Century*. London: Routledge.

Banister, D. 2007b. The sustainable mobility paradigm. *Transport Policy*, 15 (2), 73–80.

Carlowitz, H.C. von 1713. *Sylviculturaoeconomica oder Haußwirtschaftliche Nachricht und naturgemäße Anwendung zur wilden Baum-Zucht*. Leipzig: Braun (Reprint: TU Bergakademie Freiberg, Universitätsbibliothek, Nr. 135, Freiberg, 2000).

Dicken, P. 1998. *Global Shift. The Internationalization of Economic Activities*. 3rd edition. London: Guilford Press.

Earth Summit 2012. *Institutional Framework for Sustainable Development*. Available at: http://www.earthsummit2012.org/conference/themes/institutional-framework-for-sd [accessed: 09 October 2012].

EEA – European Environmental Agency 2011. *Laying the foundations for greener transport — TERM 2011: transport indicators tracking progress towards environmental targets in Europe* (EEA Report No 7/2011). Available at: http://www.eea.europa.eu/publications/foundations-for-greener-transport [accessed: 30 June 2013].

European Commission 2009. *A sustainable future for transport*. Available at: http://ec.europa.eu/transport/publications/doc/2009_future_of_transport_en.pdf [accessed: 18 September 2012].

European Commission 2011. *EU transport in figures: Statistical Pocketbook, 2011.* Available at: http://ec.europa.eu/transport/publications/statistics/doc/2011/pocketbook2011.pdf [accessed: 18 September 2012].
Eurostat 2007. *Analysis of national sets of indicators used in the National Reform Programmes and Sustainable Development Strategies.* Luxembourg.
Gaffron, P. 2012. Urban transport, environmental justice and human daily activity patterns. *Transport Policy*, 20, 114–27.
Gerike, R. 2007. *How to Make Sustainable Transportation a Reality – the Development of Three Constitutive Task Fields for Transportation.* München: Oekom.
Grieco, M. 2006. Accessibility, mobility and connectivity: The changing frontiers of everyday routine. Special issue of *European Spatial Research and Policy*, 13 (1): 1360–80.
Gudmundsson, H. 2004. Sustainable Transport and Performance Indicators. *Issues in Environmental Science and Technology*, Jg. 20.
Hyllenius, P., Smidfelt, L., Rosqvist, S., Haustein, J., Welsch, M., Carreno, T. and Rye, T. 2009. MaxSumo: Guidance on how to plan, monitor and evaluate mobility projects. Brussels: MAX – Successful Travel Awareness Campaigns and Mobility Management Strategies. Available at: http://www.max-success.eu/ [accessed: 05 December 2012].
IEA – International Energy Agency 2012. *Transport, Energy and CO2: Moving Toward Sustainability.* Available at: http://www.iea.org and http://www.iea.org/topics/transport/ [accessed: 05 December 2012].
Joumard, R. and Gudmundsson, H. 2010. *Indicators of environmental sustainability in transport: an interdisciplinary approach to methods* (Report No. 282). INRETS.
Kaufmann, V., Bergmann, M.M. and Joye, D. 2004. Motility: Mobility as capital. *International Journal of Urban and Regional Research*, 28 (4), 745–65.
Lang, D. J., Wiek, A., Bergmann, M., Stauffacher, M., Martens, P., Moll, P., Swilling, M. and Thomas, C. 2011. Transdisciplinary research in sustainability science: practice, principles, and challenges. *Sustainability Science*, 7 (Suppl. 1): 25–43.
Litman, T. 2006. Issues in Sustainable Transportation. *International Journal of Global Energy Issues*, vol. 6, no. 4, 331–47. Victoria Transport Policy Institute. Available at: http://www.vtpi.org/sus_iss.pdf [accessed: 25 April 2010].
Litman, T. 2008. *Sustainable Transportation Indicators. A Recommended Research Program For Developing Sustainable Transportation Indicators and Data.* Available at: http://www.vtpi.org/ [accessed: 25 April 2010].
OECD 2008. *2008 Annual Report on Sustainable Development Work in the OECD.* Available at: http://www.oecd.org/dataoecd/61/32/42177377.pdf [accessed: 25 April 2010].
Rauschmayer, F., Omann, I. and Frühmann, J. 2011. *Sustainable Development: Capabilities, Needs, and Well-being.* Routledge Studies in Ecological Economics 9. London: Routledge.

Schneider, N.F. and Collet, B. 2010. *Mobile Living Across Europe II. Causes and Consequences of Job-related Spatial Mobility in Cross-national Comparison.* Opladen and Farmington Hills: Barbara Budrich Publishers.

Skinner, I., van Essen, H., Smokers, R. and Hill, N. 2010. *Towards the de-carbonisation of EU's transport sector by 2050. Final report produced under the contract ENV.C.3/SER/2008/0053 between European Commission Directorate-General Environment and AEA Technology.* Available at: www.eutransportghg2050.eu [accessed: 05 December 2012].

United Nations 1992. *Report of the United Nations on Environment and Development.* Rio de Janeiro, 3–14 June 1992. Available at: http://www.unep.org/ [accessed: 05 December 2012].

UNWTO – World Tourism Organization 2012. *Tourism Highlights, Edition 2012.* Available at: http://mkt.unwto.org/en/barometer [accessed: 05 December 2012].

Urry, J. 2000. *Sociology Beyond Societies: Mobilities for the Twenty First Century.* London and New York: Routledge.

Urry, J. 2007. *Mobilities*. Cambridge: Polity Press.

WCED – World Commission on Environment and Development 1987. *Our Common Future.* Available at: http://www.un-documents.net/wced-ocf.htm [accessed: 05 December 2012].

Wiederkehr, P., Gilbert, R., Crist, P. and Caïd, N. 2004. *Environmentally Sustainable Transport (EST) Concept, Goal, and Strategy – The OECD's EST Project* (EJTIR, 4, no. 1 (2004)). Paris: 11–25.

# PART I
# Activity Patterns in Different Contexts

# Chapter 1
# The Role of Street Network Connectivity and Access to Everyday Facilities in Shaping Everyday Walking and Cycling in English Cities

Tim Jones, Dave Horton, Caroline Mullen, Colin Pooley, Griet Scheldeman, Emanuele Strano and Miles Tight

## Walking and cycling's contribution to sustainable cities

Walking and cycling are widely recognized as important forms of urban sustainable mobility because they offer benefits both to individuals and wider society. For the individual, they provide opportunities for cheap independent personal mobility, particularly among younger and older people and those who cannot or do not wish to drive, and they are also beneficial to personal health and mental wellbeing. At a societal level walking and cycling can contribute to making cities better places to live by replacing energy intensive carbon-based forms of private mobility and by reducing pollution and road danger. They can also free up land currently occupied by 'traffic architecture' to create more creative and productive use of space.

The majority of cities in Britain outside of London are dominated by private car-based mobility. Many short journeys continue to be undertaken by car despite a significant proportion of these being amenable to walking and cycling. For example, in Britain around one in five journeys are up to one mile in length (a distance easily walked) but around 20 per cent of these journeys are made by car. Furthermore, another one in five journeys is one to two miles in length (a distance amenable to cycling) but around 60 per cent of these journeys are made by car (National Travel Survey 2011a). Unfortunately, despite a couple of examples that are often cited as 'bucking the trend' such as central London and a few areas in a handful of other cities, the overall trends do not suggest that walking and cycling are flourishing across Britain. In fact quite the opposite appears to have happened over the past decade. During the period 1995 to 2010, for example, the proportion of all journeys made in Britain on foot continued to decline from 26.9 per cent to 21.8 per cent while the proportion of journeys made by cycle remained below 2 per cent – 1.7 per cent in 1995 and 1.5 per cent in 2010 (National Travel Survey 2011b). Despite the decline in walking in Britain the overall level is not too dissimilar to most other European countries (Basset et al. 2008).

The proportion of journeys made by cycle, however, remains stubbornly low in comparison to other countries in northern Europe such as the Netherlands (27 per cent), Denmark (18 per cent), Finland (11 per cent), Sweden (10 per cent), Germany (10 per cent) and Belgium (8 per cent) (Pucher and Buehler 2008).

Walking and cycling have been promoted by successive governments over this period as an integral part of UK transport policy. More recently they have also become important components of public health promotion strategies under the moniker 'active travel'. This is because they offer the potential to build in moderate levels of physical exercise if maintained as part of a daily travel routine and can help prevent diseases linked to sedentary behaviour. But while current UK physical activity guidelines suggest that adults aged between 16 and 65 should be at least moderately active for 150 minutes per week (Department of Health 2011) the evidence shows that walking and cycling are becoming less significant in people achieving this goal. Between 1995 and 2010, for example, the actual time spent walking per annum declined from 73 to 59 hours for walking and from 6 to 5 hours for cycling (Department for Transport 2011). The response from policy makers has been to include, within policy discourse, reference to walking and cycling as a means of promoting sustainable mobility *and* public health. For example, a joint policy statement issued by the UK Department for Transport and the Department of Health in 2010 stated:

> More walking and cycling … for short journeys has benefits for individuals in terms of their health – they are more likely to achieve a healthy weight and to have better mental well-being. There are benefits for communities too with safer and more pleasant streets, better air quality and lower carbon emissions, and reduced congestion (Department for Transport and Department of Health 2010: 4).

This chapter focuses on the built environment and its relationship with walking and cycling and why improving street connectivity and access to facilities on its own is *unlikely to be sufficient* in encouraging more widespread use. To provide a background, the first section begins by introducing the key facets of the built environment that are purported to shape walking and cycling and how recent research has attempted to understand their influence. Then a large-scale multi-method study is introduced that sought to understand travel decision making in relation to walking and cycling in English cities. First, the innovative approach that was used to generate quantitative spatial data as part of that study is explained followed by the results of analysis testing the relationship with walking and cycling. The chapter then draws on evidence from the qualitative component of the study to reveal the complex set of factors that discourage walking and cycling for short urban journeys. The chapter concludes with a brief summary of the implications for policy makers and the measures that should be taken if walking and cycling are to play a more instrumental role in achieving sustainable mobility in English cities.

## The role of the built environment in shaping walking and cycling

The built environment is considered by policy makers to be a key influence of walking and cycling. Urban planning is predicated on the assumption that areas that incorporate a mix of land uses along a highly connected street network are more likely to encourage more walking and cycling and there is some evidence to support this (Ewing and Cevero 2001, 2010, Handy et al. 2006, Mitra et al. 2010, Page et al. 2010). Since the 1990s the European Commission has been an influential advocate of compact urban form that contains urban development and makes use of brownfield development within existing urban areas close to everyday activities (Commission of the European Communities 1990). This approach to urban development is also reflected in the UK Urban Policy White Paper of 1999 (Department of the Environment, Transport and the Regions 1999) which highlights the benefits to quality of life gained through the promotion of dense, mixed-use developments within existing urban areas because of proximity to local facilities which are required on a daily basis.

The configuration of the street network in terms of its connectivity (or permeability) linking activities is thought to have an important influence on walking and cycling (Gehl and Gemzøe 2004). The Victoria Transport Policy Institute (2012) defines connectivity as 'the directness of links and the density of connections in path or road network'. Highly connected road networks serve to increase route options, decrease travel distance and allow more direct travel between destinations. Marshall (2005: 13) describes streets as the '… fundamental building block of urban structure … "movement space" constituted by streets forms the essential connective tissue of public urban space'. Current UK design guidance reinforces the importance of ensuring that the layout of new street networks and the retrofitting of old maximizes connectivity:

> Street networks should, in general, be connected. Connected, or 'permeable', networks encourage walking and cycling, and make places easier to navigate through. They also lead to a more even spread of motor traffic throughout the area and so avoid the need for distributor roads with no frontage development (Department for Communities and Local Government 2007: 41).

Walking and cycling are more sensitive to distance because they rely on higher levels of personal energy expenditure relative to more 'passive' forms of mobility such as car driving. Providing a mix of activities along a highly connected street network within a densely populated area (as opposed to mono-functional, segregated and more sparsely populated areas) could therefore theoretically increase the possibility of more people making a greater number of journeys to access everyday facilities on foot and by cycle. Taking the home as an important starting point for everyday journeys, it follows that those households located in the most highly connected (or 'plugged-in') areas along the street network, and that have good access to everyday activities in terms of number and diversity, are more likely to walk and cycle.

While it seems intuitive that connectivity (or permeability) of the street network is a key influence on walking and cycling, surprisingly there is little empirical research supporting this hypothesis. The evidence that does exist tends to support the hypothesis that built form affects levels of walking and cycling (Saelens et al. 2003) and that mixed land use is the most important characteristic of the built environment because of increased opportunities and convenience for accessing destinations (Badland and Schofield 2005). However, the evidence for walking is more compelling than for cycling (and often the two are combined under 'non-motorized' travel therefore masking any clear distinction).

Typical approaches to investigating the link between built environment characteristics and walking and cycling include selecting so-called 'walkable' and 'cyclable' neighbourhoods and then conducting social surveys with residents within those neighbourhoods to compare travel behaviour (Van Dyck et al. 2009), employing direct measures of connectivity around specific locations such as schools (Giles-Corti et al. 2011), or assessing facility provision within a specific range of the home (Hillman et al. 1976, Farthing et al. 1996) and testing for associations with walking and cycling. However, there is general ambiguity and inconsistency in how to actually measure connectivity or what level of connectivity is desirable for walking and cycling to flourish. This makes it difficult for valid comparison to be made across separate studies or geographical areas (Moudon and Lee 2003). Existing studies are largely focused on cases in North American and Australia and few have examined the different spatial context across Europe. These issues have led Dill (2004) to call for the advancement of techniques for measuring built environment factors such as connectivity and land use mix to enable more robust conclusions to be made about their apparent influence on walking and cycling across different geographical domains. The next section discusses a large-scale study of walking and cycling in England which included the use of innovative methods of measuring street connectivity and land use to investigate their relationship with walking and cycling.

## Investigating the relationship between land use activity/connectivity and walking and cycling: the English Understanding Walking and Cycling study

The Understanding Walking study was a large UK Research Council funded project undertaken during the period 2008 to 2011, which aimed to investigate the ways in which households incorporate (or fail to incorporate) walking and cycling into routine everyday travel (for example to work, shops and so on) and the decision-making processes that lead to specific travel behavioural outcomes. The overall approach used departed from the dominant approach to travel behaviour research which typically applies micro-economic theory to travel choice modelling (see Gärling 1998, Walker 2006 and Burnett 2007 for discussion on limitations and opportunities) and instead used a mix of quantitative and qualitative methods to examine the micro-scale complexity of household decision making and associated travel strategies.

Four English case study cities were selected (Lancaster, Leeds, Leicester and Worcester) to represent a range of conditions in terms of geography and demography that exist across England – brief details are provided in Box 1.1. The mixed methods applied across these four areas included a social survey of a large sample of households probing experience of and attitudes towards walking and cycling; spatial analysis of the built environment using indices of connectivity and analysis of land uses; in-depth interviews with a selection of householders drawn from the survey probing attitudes to walking and cycling and the reasons why people chose particular modes of travel; interviews with participants whilst they were making a 'usual' journey on foot or by cycle (accompanied walk/ride-along interviews) focusing particularly on motivations, route selection and on the experience of the journey; and, finally, more immersive ethnographic studies with a selection of households over an extended period of three months to observe and understand the nature of everyday journeys within a community using a combination of research tools including home-based interviews and interviews whilst moving around the city, mobility inventories, observations, mapping exercises and community participation. More detailed information about the approach and methods used is available in Jones et al. (2009) and Pooley et al. (2011, 2013).

### Box 1.1 The case study cities

**Lancaster** located in the north west of England is a small compact city with outer areas extending no more than 5 km from the urban core. The population of the city is around 46,000 (Census, 2001) but the combined non-metropolitan population is around 134,000 when nearby satellite towns of Morecambe, Heysham and small villages are taken into account. The population density is only 2.3 people per hectare (Census 2001) within the larger Lancaster administrative boundary (57,586 hectares) because it incorporates a large proportion of agricultural and undeveloped land within the non-metropolitan area. The River Lune river plain separates Lancaster and Morecambe and Heysham to the north some 12 km away. The topography is relatively flat although land rises steeply and uniformly immediately to the east of Lancaster.

**Leeds** located in the north of England is a large city with a continuously built up area extending up to 12 km in each direction from the city centre. The total population of the City of Leeds is around 715,000 (Census 2001) spread across an area of 55,173 hectares giving a population density of 13 people per hectare. The central area of Leeds is located on the River Aire and contained within the Leeds Inner Ring Road. The topography of the city centre is such that the land rises away from the flat River Aire valley and is generally undulating with some steep gradients – the city centre lies roughly at about 63 m above sea level and rises to 340 m to the west at Ilkley Moor.

**Leicester** located in the East Midlands is a relatively compact city and most of its urban area is located within the ring-road some 5 km radius from the city centre although some suburban development extends a further 2–3 km beyond the ring-road to the border with the green belt and countryside. The population is around 280,000 (Census 2001) living across a built up area of some 7,309 hectares. The population density stands at around 38 people per hectare and is higher than the typical English average. The River Soar and Grand Union Canal flow from south to north through the city centre and the topography of the urban area is generally flat.

**Worcester** located in the south Midlands is a medium-sized, relatively compact city with outer areas no more than about 5 km from the city centre. The total population is around 93,000 (Census 2001) housed in some 40,000 dwellings across an area of 3,328 hectares giving a population density of approximately 28 people per hectare. Worcester is located along the River Severn that cuts north–south through the city with its historic and commercial centre and most of its urban area lying to the east. The topography of the city is relatively flat but there is a gentle rise in elevation mainly to the east of the river.

The remainder of this chapter draws on combined data from the social survey and spatial analysis together with elements of the qualitative inquiry. The approach used for spatial analyses and associations with walking and cycling is discussed in the next section and insights from the qualitative inquiry follow that. Together these data provide compelling evidence of the multiple factors that shape walking and cycling and the significance of the built environment within this mix.

*Social survey*

The postal questionnaire used in the Understanding Walking and Cycling study for the social survey of households across the four case study cities was geo-coded so that self-reported travel data could be linked with spatial analyses. A total of 15,000 households were selected randomly across the four cities with the sample stratified by social economic group using UK indices of multiple-deprivation. A single adult (aged 18+) from each selected household was asked to answer questions related to their attitudes to walking or cycling and to write in the frequency of use of different methods of travel during a typical week for five non-work related journey purposes (visiting nearby friends and relatives, leisure journeys, personal business, grocery shopping and other shopping).

A total of 1266 cases (circa 10 per cent of sample) provided complete travel data enabling these to be linked with spatial data. Responses were evenly spread throughout each of the four case study cities and were broadly representative of the total population of each town although there was a slight bias towards females and to those from higher social groups. It is also likely that those who engaged in walking and cycling were most likely to respond, although it was stressed that the survey was relevant to all irrespective of how often they walked or cycled.

### *Spatial analysis*

Spatial analysis consisted of constructing a complete street network of 'nodes' (intersections) and 'edges' (links) for each case study city using Ordnance Survey Integrated Transport Network (OS ITN) mapping and GIS software. Links where walking and cycling were not permitted (such as motorways) were excluded and local transport authority-produced walking and cycling maps were used to identify potential areas in the network that might not have appeared on the ITN layer. It should be noted that the final network model was not able to take into account the 'regulatory system' because of the various technicalities of dealing with this issue. This meant that elements such as one-way streets were ignored (that is, two-way connectivity was assumed) so the networks were likely to over-represent reality on the ground.

Each respondent's home location was plotted on a map layer. Line-based network buffers were then created around each home location using a path in all directions of 800 metres for walking and 2400 metres for cycling. These distances were based on an assumed uninterrupted average travel time of 10–15 minutes at relaxed pace with cycling covering roughly three times the distance of walking over the same travel time. A line-based approach was used in preference to circular buffers because of reported accuracy in representing what is *actually* accessible within a given space (Oliver et al. 2007).

A series of connectivity indicators were then calculated within each individual 800m and 2400m buffer and added to the dataset of social survey information. The indicators sought to represent street network connectivity within the *neighbourhood* around each respondent's home (that is, within 800 metres for walking and 2400 metres for cycling) and included *street density* (the total length of street available within each network buffer); *node density* (the total number of intersections within the network buffer); and *average node degree* (number of route options in the form of the average number of streets branching off each node). The hypothesis tested was that there is a positive relationship between measures representing connectivity of the local neighbourhood street network and the frequency of walking and cycling.

A separate set of measures was also generated to indicate level of connectivity of a respondent's home location in relation to the *entire city street network* and were derived using a technique called Multiple Centrality Assessment (MCA) (Porta et al. 2006). MCA generates a set of measures based on *topographical* properties (that is, the relation of each intersection, or 'node', to each other intersection) and *spatial* properties (that is, the distance along streets, or 'edges', between intersections) and is therefore useful for investigating walking and cycling given their sensitivity to journey distance (Scheurer and Porta 2006). Three centrality measures were used; *betweeness*, *closeness* and *straightness*. Each of the measures capture a different aspect of a specific location's significance in terms of 'being central' in relation to all other points in the network. Originating from quantitative sociology (Freeman 1977, Sabidussi 1966) *betweeness* centrality indicates the level of strategic importance of a location within the network. Those nodes with higher strategic importance are those that are traversed by many of the shortest paths connecting pairs of nodes. *Closeness* centrality indicates the extent to which a node is near to all other nodes along the shortest paths. Finally, *straightness* centrality indicates the extent to which the connecting route between nodes deviates from the virtual straight route or to put it another way, the level of directness between two nodes. Measures were generated by taking both the average and sum of values within each network buffer around each respondent's home.

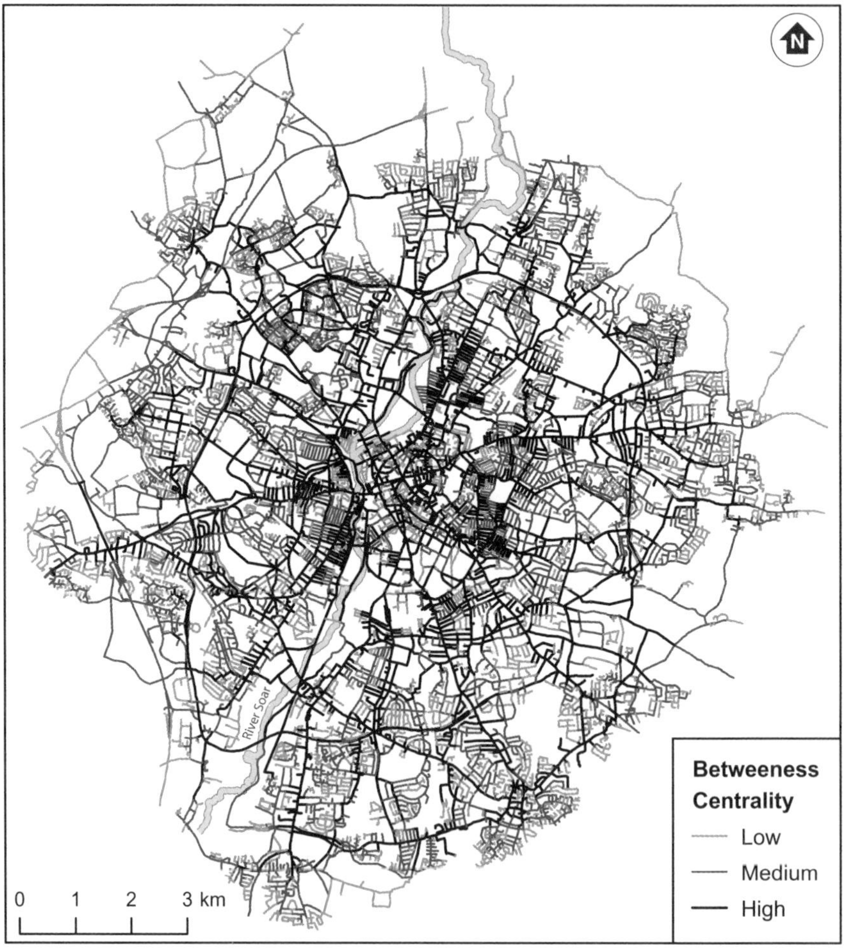


**Figure 1.1 Example of betweeness centrality map of Leicester**

Centrality measures for a city street network can be visualized with darker lines indicating areas of high centrality (or connectedness), and lighter lines showing areas of low centrality across the entire network (see Figure 1.1). Closer inspection of the betweeness centrality maps produced for each of the four areas demonstrated that the most central locations are main roads including arterial routes linking the city centre and suburbs – typically the routes that experience the highest motor-traffic flows.

The hypothesis tested was that there is a positive relationship between 'being central' in the entire city street network and the frequency of walking and cycling. It should be noted that the 'global' centrality measures are specific to each case study city street network and meaningful comparisons cannot be made between cities because the data is non-normalized. However, it is reasonable to assume that if results for each measure within each individual city point in the same direction (and similarly for local neighbourhood measures), then they corroborate each other and firm conclusions can be made about the strength and direction of relationship between connectivity and walking and cycling behaviour.

As well as the measures of connectivity an assessment was also made of the prevalence and diversity of local everyday activities within each buffer. Using Ordnance Survey Base Function categories from OS MasterMap Address Layer 2 data (Ordnance Survey 2011) 10 typologies representing common activities were identified (finance, post office, non-food retail, food retail, catering, informal recreation (for example parks), formal recreation (for example sports centres), health, education and general commercial). Using GIS software, these were overlaid on two separate maps containing separate buffers for walking (800m) and cycling (2400m). As well as an *absolute count of activities*, an entropy index was applied to provide a measure of *diversity and evenness of activities* within each buffer similar to that which an ecologist might use to measure the biodiversity of a specific habitat. The hypothesis tested was that there is a positive relationship between the prevalence and diversity of local activities in the local neighbourhood and the frequency of walking and cycling.

A summary of all of the variables included in the analysis and their definition is shown in Table 1.1.

Evaluation of the relationship between the measures of connectivity and land use and reported levels of walking and cycling (using Pearson's r) is summarized in Tables 1.2 to 1.5 below with statistically significant correlations shown in bold. Correlation analysis between local neighbourhood measures and reported frequency of *walking* for non-work related journeys shows a positive correlation but this relationship is weak (Table 1.2). The prevalence of a range of activities and the actual street density are particularly associated with a higher frequency of walking, while the diversity of activities, node degree and node density provide a less conclusive relationship because there are not clear associations across all four cities. Analysis of the association between local spatial indices estimated within a 2400 metre network buffer and reported frequency of *cycling* for non-work related journeys shows that there is no correlation between any of the local spatial measures and cycling (Table 1.3). This suggests that the characteristics of the local neighbourhood street network in terms of connectivity and presence/diversity of local activities could be more important for walking than for cycling.

**Table 1.1 Summary of variables measures within each buffer**

| | |
|---|---|
| **Global measures of connectivity** | |
| Closeness centrality ($C^{C}_{GLOB}$) | Defined as the extent to which a node is near to all the other nodes along the shortest paths, or to put it another way, the shortest path length between two nodes throughout all possible paths. |
| Betweeness centrality ($C^{B}_{GLOB}$) | Represents the average proportion of paths between two nodes that traverse the node in question. A node is central if it lies between many other nodes in the sense that it is traversed by many of the shortest paths connecting couples of nodes. |
| Straightness centrality ($C^{S}_{GLOB}$) | Represents efficiency in communication between two nodes and is equal to the inverse of the shortest path length. Captures the extent to which the connecting route between nodes deviates from the virtual straight route. |
| **Local measures of connectivity** | |
| Street density | Total length of streets (edges) within the network buffer. |
| Street length (average) | The average length of street (edges) between two nodes within the network buffer. |
| Street density | Total length of streets (edges) within the network buffer. |
| Node density | The total number of intersections (nodes) within the network buffer. |
| Node degree (average) | The average number of streets (edges) per intersection (node) within the network buffer. |
| **Local measures of land use activity** | |
| Land use activity categories | Aggregated from Ordnance Survey Base Function categories in Address Layer 2 data.<br>Ten classifications were derived:<br>1. Finance<br>2. Post Office<br>3. Non-food retail<br>4. Food retail<br>5. Catering<br>6. Informal recreation<br>7. Formal recreation<br>8. Health<br>9. Education<br>10. General commercial |
| Activity prevalence | The total number of land use activities (indicated using the ten land use activity categories) within the network buffer. |
| Activity diversity | The diversity of land use activities (using the frequency of each of the ten land use categories) within the network buffer derived using the Shannon Index. |

**Table 1.2 Association between local neighbourhood measures and reported frequency of walking for non-work related journeys**

| | **Land use activity** | | **Street connectivity** | | | |
|---|---|---|---|---|---|---|
| | Activity diversity | Activity prevalence | Node density | Node degree (ave. metres) | Street density | Street length (ave. metres) |
| ALL | **+.10**** | **+.19**** | **+.09**** | **+.07*** | **+.14**** | –.05 |
| Lancaster | +.02 | **+.19**** | +.09 | **+.12*** | **+.13*** | –.08 |
| Leeds | +.03 | **+.18**** | +.09 | **+.14*** | **+.12*** | –.04 |
| Leicester | **+.15*** | **+.21**** | **+.19**** | +.06 | **+.21**** | –.10 |
| Worcester | **+.14*** | **+.18**** | +.07 | +.05 | **+.14*** | +.04 |

*Notes:*

Measures calculated within an 800m line based buffer.

Significance level: *=p<.01; **=p<.001

**Table 1.3 Association between local neighbourhood measures and reported frequency of cycling for non-work related journeys**

| | **Land use activity** | | **Street connectivity** | | | |
|---|---|---|---|---|---|---|
| | Activity diversity | Activity prevalence | Node density | Node degree (ave. metres) | Street density | Street length (ave. metres) |
| ALL | +.01 | +.03 | –.02 | +.04 | .00 | +.05 |
| Lancaster | –.08 | +.08 | +.05 | –.00 | +.07 | +.04 |
| Leeds | +.01 | +.08 | +.04 | +.01 | +.03 | –.01 |
| Leicester | +.05 | –.05 | +.02 | +.03 | +.03 | –.04 |
| Worcester | +.01 | +.05 | +.02 | –.01 | +.02 | –.01 |

*Notes:*

Measures calculated within a 2400m line based buffer.

Significance level: *=p<.01; **=p<.001

Further analysis of the correlation between centrality measures estimated within an 800 metre network buffer and reported frequency of *walking* for non-work related purposes also indicates a positive but weak correlation between the majority of global spatial indices and frequency of walking for all of the cities apart from Leeds where no relationship was found between any of the measures (Table 1.4). Those living nearest to the city centre, as measured by the index of closeness, are most likely to walk. The result of correlation analysis between centrality measures estimated within a 2400 metre network buffer and reported frequency of *cycling* for non-work related journeys again shows that there is no correlation between any of the global spatial measures and cycling (Table 1.5).

**Table 1.4 Association between centrality measures based on home location within entire city street network and reported walking frequency for non-work related journeys**

| | Sum betweeness | Average betweeness | Sum closeness | Average closeness | Sum straightness | Average straightness |
|---|---|---|---|---|---|---|
| Lancaster | **+.11*** | +.03 | **+.15**** | **+.11*** | **+.12*** | **+.15**** |
| Leeds | +.13 | +.02 | +.16 | +.10 | +.16 | +.08 |
| Leicester | **+.23**** | **+.19**** | **+.22**** | **+.24**** | **+.19**** | +.04 |
| Worcester | **+.15**** | **+.11*** | **+.12*** | **+.13*** | +.09 | +.09 |

*Notes:*

Measures calculated within an 800m line based buffer.

Significance level: *=p<.01; **=p<.001

**Table 1.5 Association between centrality measures based on home location within entire city street network and reported cycling frequency for non-work related journeys**

| | Sum betweeness | Average betweeness | Sum closeness | Average closeness | Sum straightness | Average straightness |
|---|---|---|---|---|---|---|
| Lancaster | +.08 | +.06 | +.06 | +.06 | +.03 | –.05 |
| Leeds | +.03 | –.01 | +.04 | +.03 | +.03 | +.05 |
| Leicester | +.03 | +.02 | +.03 | +.03 | +.03 | +.06 |
| Worcester | +.05 | +.05 | +.01 | –.04 | +.02 | –.04 |

*Notes:*

Measures calculated within a 2400m line based buffer.

Significance level: *=p<.01; **=p<.001

The findings from this element of the Understanding Walking and Cycling research suggest that the connectivity of the street network and the availability of everyday activities within walking and cycling distance of the home may have some influence, but are insufficient on their own to encourage walking, and in particular, cycling. However, this does not suggest that they are of no consequence, but rather that they are not the dominant factors shaping everyday travel decisions at least within the case study areas investigated. The configuration of the built environment provides the foundations for walking and cycling to flourish, but as we reveal in the subsequent section, other factors militate against walking or cycling for everyday travel.

## Why street network connectivity and access to facilities are insufficient in encouraging walking and cycling: insights from qualitative analysis

The qualitative data accumulated during the Understanding Walking and Cycling study provided the opportunity to add possible explanation to the associations revealed in the previous section. Data were generated from (a) interviews with 10 households in each case study area (b) mobile interviews with 10 individuals whilst they were making a typical routine journey either on foot or by cycle again in each case study area (c) five in-depth ethnographic studies with households in selected neighbourhoods in each of the four case study areas chosen to represent a diverse demographic. A total of 250 separate transcripts were generated from the 100 households taking part in this part of the study. The combination of data generated from the quantitative and qualitative approaches therefore provided a powerful resource for examining contemporary attitudes to walking and cycling and the circumstances that encourage or discourage their practice. Unfortunately, there is not the space here to provide a full account of the specific geographical, social and cultural context within which data were generated and the implications for promoting walking and cycling. Readers interested in the distinct differences between areas should refer to Pooley et al. (2013a). What follows is a general account of key findings relevant across *all* of the different geographical, social and cultural contexts.

First, it is worth pointing out that there was distinction between the extent to which people across all areas were able to articulate their opinions or experiences of walking and of cycling. Walking is something that the majority of people invariably do even if they do not walk very far or very often. In fact, it is barely visible, that is to say, it is something people do almost unconsciously without giving it much thought (Pooley et al. 2013a). Conversely, and somewhat paradoxically, most people do not cycle but do have something to say about it whether in a negative sense (for example, the problematic pavement cyclist was a common topic of discussion) or positive in terms of recognition of the benefits to health and environment. When conducting household interviews it was common for participants to struggle to find anything remarkable to say about walking precisely because it is such a mundane taken-for-granted activity. The mobile interviews and ethnographies were more successful in this respect because of the obvious natural prompts that were encountered whilst moving around the city with participants. For example, participants were able to point out specific examples of affordances and challenges that they encountered along their routes; or they were observed behaving in certain ways in particular traffic situations and were then able to articulate their reasons when prompted.

### *Household and family factors*

What emerged from the combination of questionnaire survey, in-depth interviews and ethnography was that, although attitudes to walking and cycling were mostly positive or at least indifferent, people who would like to engage in walking and (particularly) cycling did not do so for a combination of factors. A significant reason was the difficulty households perceived fitting walking and cycling into everyday routines, and this particularly applied to households with children (see Box 1.2). Walking was often incompatible with the way that multi-person households lived their everyday lives and was rarely simple even in circumstances where there was the potential to replace car journeys. Careful planning was required in terms of assembling shoes, jackets, bags, pushchairs and other accessories to travel, particularly where journeys were complex or multi-purpose and during certain times of the day or when the weather was poor. Cycling was rarely considered despite its extended range over the same travel time when compared to walking (for reasons that we will explain later), and when it was, this was mostly in the context of going out for leisure journeys along off-road tracks away from traffic.

**Box 1.2 Household constraints on walking and cycling**

With the demands of family and work and everything there's not much time or energy [for walking or cycling] (Percy, Worcester, go-along).

I think time between activities makes a difference, even though it may be a short journey, we probably wouldn't walk or cycle because we've only got so much time to get from one place to another (Pete, Worcester, mobile interview).

Children influence walking routes (both through wanting (insisting) to go a particular way and through parents wanting to take them a nicer/safer/less polluted way). … Distance is important for walking, especially if with kid or if carrying stuff (Hailey, Leeds, mobile interview).

Usually I go with the car because of convenience, less time, because sometimes J [age 3] is tired when I pick him up from nursery and I would have to carry him, and I have my books as well, and when there's two of them … In the week [I use the car] for two days a week, at some point I might even try just walk with the kids, but it's usually because with the two kids they have different energies, and R runs and J is a bit more like staying here and hanging round here and there, so that creates some kind of tension and also I'm on pressure to get on time to work, then it's really much more convenient to just strap them on the seats and take them and leave them and that's it. Apart from that I would just walk (Don, Lancaster, interview).

*Perceptions of normality*

For most households, for everyday travel, it was routine to simply 'jump in the car'. This was particularly the case for households located in more suburban areas where access to facilities of choice meant complex, multi-purpose journeys with other family members. However, even in areas where local facilities were available and the street network was seemingly well connected, often the car was the taken for granted 'normal' means of moving around in that it offered perceived advantages of speed, comfort, security and the ability to move around as a family unit. Among the majority of participants, whether they walked or cycled to a greater or lesser degree, there was the perception that walking and cycling are simply no longer customary ways of moving around British cities as the car has come to dominate everyday travel to the extent that walking and cycling are now seen as abnormal and are stigmatized (see Box 1.3).

**Box 1.3 (Ab)Normality/stigma associated with walking and cycling**

You don't see a lot of people walking around my neighbourhood (Thomas, Worcester, ethnography).

People still assume that there's something wrong with you if you don't drive. … I would walk a lot more if I could. Often I don't have the right shoes with me (Lara and Bob, Leeds, interview).

The whole thing with transport and not having a car, I do feel like a second class citizen, there's definitely a sense that as a pedestrian and a cyclist you are definitely second class citizens (Jim, Lancaster, interview).

Other people might think I'm trying to save money … So it's almost like I should be ashamed to cycle – it's because I can't afford a car (Dhi, Leicester, ethnography).

When the elderly are riding a bicycle, the young ones make fun of them. That's one part. I've seen it, I don't know if anybody's seen it, when elderly people are riding bicycle, young ones you know, make fun of it (Devaraj, Leicester, ethnography).

*Physical environment*

While complex household circumstances including location, structure and routine were barriers to walking and cycling, as well as perceptions that they are somehow not normal vis-à-vis travel by car, the perceived nature of the physical environment also served as a barrier. This is particularly pertinent given that it was highlighted in the opening section of this chapter that there is significant potential for at least some short journeys to be made on foot and by cycle. For walking the quality of the physical infrastructure was not so much a major factor as the perceived threats posed by the social environment. So while lack of pavement space, uneven pavement slabs and poorly sited crossings were just some of many obstacles noted by researchers, participants seemed to accept the infrastructure as it currently exists often without question and without reflecting on how it could be better. Moreover, general deference by participants to motor traffic was observed whilst they were walking even in situations where it was their right to exercise priority on the highway (such as when crossing side roads). As noted earlier, walking is an activity that is rarely considered precisely because it is regarded as mundane and taken-for-granted. It is not surprising therefore that those participants did not offer critical appraisals of the conditions they encountered whilst walking in the city because certain customs have become the norm. Where concerns were raised this was often about risks posed by motor traffic and to personal safety by others and how this influenced where and when people walked (see Box 1.4). Females and older people in particular spoke about restricting their walking after dark and avoiding empty streets or varying routes to avoid areas with a reputation for being a perceived threat to personal safety. The conflicts caused by people riding bicycles on the footway and the degree of incompatibility of walkers and cyclists sharing the same space was also a key topic of discussion – a point we return to in the next section.

**Box 1.4 Constraints posed by the urban environment for walking**

*Urban infrastructure*

I don't like walking on busy roads. I hate the big lorries. It's the big vehicles that frighten me whether I'm on foot or on the bike (Jan, Leeds, interview).

Two quite dangerous roads, even though they're quite narrow. They're not like big roads like the inner city ring road, but almost as dangerous to cross, just to go and get a pint of milk (Heidi, Leeds, ethnography).

That road is awful, the pavement is very narrow, and in autumn it's covered in leaves so you slip over half the time, it's terrifying. But by car the road is fine (Linda and Paul, Lancaster, ethnography).

Whoever designed this estate didn't have pedestrians in mind. … I have to go on a narrowish path with big piles of dog dirt on it (Nancy, Leeds, interview).

*Perceptions of safety and risk*

If I want to go to the Post Office, there's one quite close but I'll take the car because I don't like walking through the estate. … I feel very vulnerable walking some places because I can't run. … Walking through alleyways can be very claustrophobic and you feel a bit scared so I tend to walk the long way round on the road which adds journey time, but it's worth it (Jen, Worcester, interview).

I know the good areas and the bad areas in the city and I always make sure I am walking with someone, or that I am walking at the right time. … I don't feel I have to be aware of anything at all when I am walking around. I pretty much know everyone in the area. It's just being familiar with faces, we're not a close knit community but I think people would look out for each other if there was a problem (Anju and Pooja, Leicester, ethnography).

I feel safer going through that street where there's a lot more people around, rather than that road where you've got the cars but you don't really have many people walking it.

… I think you've just got to be vigilant whatever you do, and not be too complacent about things. But if you let your fears of walking places get the better of you then you're never going to do anything. … In the winter months people don't come out, as they don't like to be going home in the dark. … I prefer to walk on streets with more houses because I think you feel a bit more safer when there are more things. And that's the main reason for walking this route in winter when it's dark (Steve, Leeds, ethnography).

*Pavement cycling*

I'd say you see more cyclists on the pavement [in Leicester] than you do on the road. … I can see exactly the pedestrian's point of view they don't expect to see a cyclist (Richard, Leicester, mobile interview).

I've always felt as a pedestrian, much more threatened by cyclists. … I have a fear because a lot of the cyclists don't seem to obey rules of the road, especially when it comes to traffic lights and chopping and changing between being on the road and being on the pavement and so I've almost been hit far more times with a bicycle [than by a car] (Tim, Leicester, mobile interview).

If you want to be really safe and that then you'd have to ride on the pavement (Neela, Leicester, interview).

The key narrative around cycling was that conditions for its everyday practice as a means of moving around the city were poor. While a small proportion of participants who cycled themselves seemed to thrive under these circumstances, or had learned to cope with them, the vast majority of the research participants simply refuse to cycle because of their fear of riding in an environment dominated by motorized traffic. There was the perception that everyday urban cycling in Britain is performed only by the most committed and resilient cyclists who are comfortable, or actually thrive on, interacting with faster and more dominant motor traffic on British roads. Paradoxically, it was typical for cyclists to acknowledge within their narratives the confidence and skills they possessed that enabled them to employ the necessary tactics and strategies to negotiate and 'survive' the current highway system (see Box 1.5). Less confident riders often revealed how they created their own space, often along pavements, to avoid sharing roads with motor traffic (to the ire of some pedestrians as mentioned earlier). They recognized that this action could pose problems for pedestrians but felt they often had little option given the perceived unsafe conditions on the road. A significant proportion of (mainly female) participants had at some point in their lives experimented with cycling but had given this up after bad experiences involving other road users. So while the potential for cycling in cities is arguably greater because of its relative speed and ability to carry luggage (and children), it largely remains unfulfilled because of public anxiety of riding under current conditions regardless of proximity of and level of connectivity to local facilities.

### Box 1.5 Constraints posed by the urban environment for cycling

I feel that the cyclists who are the safest are the ones who make themselves big and 'I'm here and I'm keeping up with the traffic and I know what I'm doing', and you make yourself obvious (Audrey Lancaster, ethnography).

I absolutely love my cycle to work. I like cycling anyway and I have always wanted to be in a position to be able to cycle to work and now I am … I quite like the challenges of wrestling with the traffic … In winter, in the dark, you have to be much more aware; not every car driver is bike aware. I need my lights in good working order and protective clothing for the weather … I'd like to use the car less … Riding the bike to work is beginning to make me think about the greater capacity to use the bike for other things … It's quite scary on the ring road on a bike – people come down at 60, 70, 80 mph (Katie, Leeds, interview).

I used to cycle, but because I got knocked off twice early on, that puts you off … Cars always want to overtake you … they don't feel that you're another vehicle on the road; they just see you as a pedestrian on the road, and an inconvenience (Andy, Worcester, mobile interview).

I used to bike, I did really enjoy biking … But once, I was biking in the road and a car beeped at me and it really put me off. I didn't do it much after that, it was one of my last times I think, because I really didn't like that. I lived in a little town and the pavements are really narrow so people don't like you biking on the pavements, but people don't like you on the road either, so I had to be in the road and I didn't like it, I don't like cars being behind me, and having to stop when cars stop and all that … I don't want to do it on these roads, there's no chance (Claire, Lancaster, interview).

I did my cycling proficiency at primary school and I took a bike with me when I first went to university in St Andrews, but I'm a complete coward when it comes to cycling on roads … I just, everybody I know who does a considerable amount of cycling at some point has been knocked off and hurt themselves in some way, and as a driver, cyclists on the road just seem so vulnerable that I just don't want to join them, and although the university cycle path is really good as far as it goes, erm, even so you can't avoid cycling on the road at some point, and if not cycling on the road then cycling on the pavement, which isn't legal and does expose pedestrians to risk, which I can't justify. So, you know, it's just not something that I'd consider (Hilary, Lancaster, interview).

I am not a happy cyclist … I gave my bike away, to someone who was going to use it and get benefit from it … I'm not comfortable in traffic, so the kind of bike riding I'd prefer to do would be away from traffic completely. I know we have lovely cycle paths. I could go to Morecambe and then even along the prom, but I would have to get there on a bike, and I couldn't face that (Anne, Lancaster, interview).

My ideal would be if it were possible, transport wise, for cycle paths to be absolutely physically removed from roads as in a proper kerb separating cyclists from traffic so that cyclists didn't have to use the pavement but weren't sharing the road with cars then cycling would definitely be an option and I'd find ways around the other inconveniences of cycling. But as I say with cyclists having to mix with traffic is just seems crazy (Holly, Lancaster, interview).

## Conclusion: shaping the environment for walking and cycling

This chapter has drawn on quantitative and qualitative data generated from the large-scale UK Understanding Walking and Cycling study to investigate decision making in relation to walking and cycling in English cities. The common assumption that simply providing activities close to homes along a highly connected street network will somehow encourage walking and cycling is a tenuous one. This does not mean that they are unimportant, indeed they are a necessary pre-requisite, but they are *insufficient* on their own insofar as there are other primary factors that inhibit walking and (particularly) cycling.

The qualitative data revealed that walking is used for at least some journeys but this is often conditional on factors such as home location in relation to other activities, family circumstances, co-ordination, timing and perceptions of the urban environment. This is also true of cycling, but as demonstrated, cycling only features in the everyday lives of a minority of the population who are mostly skilled and confident when riding in traffic, or who otherwise adopt coping strategies, such as pavement cycling, to avoid motorized traffic. Lack of association of cycling with proximity of activities or a connected street network reinforces the fact that those who currently cycle often do so regardless of urban conditions or otherwise construct their own routes to access their destination.

The qualitative data also revealed concerns about personal safety when walking or cycling on more secluded routes and conversely concerns about the dangers posed by motor traffic on busier roads. In the UK, arterial routes into towns and cities are some of the most heavily trafficked routes and are typically lined with shops and everyday facilities. The centrality maps generated from Multiple Centrality Assessment provided a clear indication that most central or significant locations along the entire street network – the example of Leicester was included in Figure 1.1 – are precisely these main strategic arterial roads and not residential back-streets. Paradoxically, over the past two decades, the focus of UK policy has been on routing cyclists along quieter residential roads and traffic-free routes alongside rivers and canals or along disused railway corridors. Our research suggests that more attention needs to be given to improving the quality of main roads and junctions for walking and cycling because of their strategic importance for connectivity when travelling within cities and importance for accessing everyday activities. Furthermore, providing dedicated space for cycling along these routes could also improve conditions for walking because cyclists who fear sharing the highway with motor traffic would be less inclined to use the pavement.

To summarize, this chapter has demonstrated that under current urban conditions in Britain, street pattern and land use factors have only a limited effect on the propensity to walk or cycle and that other constraints imposed by family and life-style, perceptions of safety and convenience, and expectations about what means of everyday travel are normal are primary factors. For walking and cycling to become more significant in everyday mobility, the conditions within which people are expected to walk and (particularly) cycle and the structures that inhibit them need to be challenged (Pooley et al. 2013b). This will require a multi-faceted approach to create safe physical environments where most people feel comfortable walking or cycling.

Spatial organization *is* an important foundation and should ensure that access to common everyday facilities is easy and the environment is supportive. This can be achieved by dedicating space for cycling segregated from both pedestrians and motor traffic along busier roads; improving the quality and maintenance of pavements; and introducing lower speed limits across residential streets and central areas of the city. Wider societal and economic changes also need to provide people with greater opportunity to walk and cycle through improved opportunity for flexible working and also family-welfare policies. For example, parents of families with small children being able to afford to reduce working hours and be less constrained by time commitments. Finally, in tandem with these measures, there is the need to promote a significant cultural shift where walking and cycling are perceived as 'normal' ways of conducting short journeys in urban areas and car driving is perceived as 'abnormal'. This will require promoting walking and cycling as something accessible to all and not dominated by super-fit or unusually committed specialists. However, it is essential that promotional activity goes hand-in-hand with improvements in urban quality and therefore experienced reality. This requires a significant shift in UK urban transport policy whereby walking and cycling is prioritized and made safer in urban areas such that it is perceived by the public as more convenient and enjoyable than using a car in the city.

## References

Badland, H. and Schofield, G. 2005. Transport, urban design, and physical activity: an evidence-based update. *Transportation Research Part D: Transport and Environment*, 10 (3), 177–96.

Bassett, D., Pucher, J., Buehler, R., Thompson, D. and Crouter, S. 2008. Walking, cycling, and obesity rates in Europe, North America, and Australia. *Journal of Physical Activity and Health*, 5, 795–814.

Burnett, P. 2007. Qualitative techniques for urban transportation, in *Transport Science and Technology*, edited by K.G. Gouias. Oxford: Elsevier.

Commission of the European Communities. 1990. *Green Paper on the Urban Environment – Communication to the Council and Parliament COM(90) 218 (1990)*. Brussels.

Department for Communities and Local Government (DCLG) 2007. *Manual for Streets*. [Online]. Available at: http://www.communities.gov.uk/publications/planningandbuilding/manualforstreets [accessed: 27 July 2012].

Department of the Environment, Transport and the Regions (DETR) 1999. *Towards an urban renaissance: final report of the Urban Task Force*. London: E & FN Spon.

Department of Health, Physical Activity, Health Improvement and Protection 2011. *Start Active, Stay Active: A report on physical activity from the four home countries' Chief Medical Officers*. London: Department of Health.

Department for Transport (DfT) 2011. *National Travel Survey Table NTS0310: Average time spent travelling by main mode, Great Britain, since 1995/97*. [Online]. Available at: http://www.dft.gov.uk/statistics/tables/nts0310/ [accessed: 10 May 2012].

Department for Transport (DfT) and Department of Health (DoH) 2010. *Active Travel Strategy*. [Online]. Available at: http://www.dh.gov.uk/prod_consum_dh/groups/dh_digitalassets/documents/digitalasset/dh_113104.pdf [accessed: 27 July 2012].

Dill, J. 2004. Measuring Network Connectivity for Bicycling and Walking. *TRB 2004 Annual Meeting CD-ROM*.

Ewing, R. and Cervero, R. 2001. Travel and the built environment: a synthesis. *Transportation Research Record*, 1780, 87–114.

Ewing, R. and Cervero, R. 2010. Travel and the built environment. *Journal of the American Planning Association*, 76, 265–94.

Farthing, S., Winter, J. and Coombes, T. 1996. Travel behaviour and local accessibility to services and facilities, in *The Compact City: A Sustainable Urban Form?*, edited by M. Jenks, E. Burton and K. Williams. London: E & FN Spon.

Freeman, L.C. 1977. A set of measures of centrality based on betweeness. *Sociometry*, 40, 35–41.

Gärling, T. 1998. Behavioural assumptions overlooked in travel choice modeling, in *Modelling Transport* (second edition), edited by J. de D. Ortuzar and L.G. Willumsen. Chichester: Wiley.

Gehl, J. and Gemzøe, L. 2004. *Public Spaces – Public Life*. Copenhagen: Danish Architectural Press.

Giles-Corti, B., Wood, G., Pikora, T., Learnihan, V., Bulsara, M., VanNiel, K., Timperio, A., McCormack, G. and Villanueva, K. 2011. School site and the potential to walk to school: The impact of street connectivity and traffic exposure in school neighborhoods. *Health & Place*, 17 (2), 545–50.

Handy, S., Cao, X. and Mokhtarian, P. 2006. Self-selection in the relationship between the built environment and walking: evidence from North California. *Journal of the American Planning Association*, 72 (1), 55–74.

Hillman, M., Whalley, A. and Henderson, I. 1976. *Transport Realities and Planning Policy: Studies of Friction and Freedom in Daily Travel.* London: Political and Economic Planning.

Jones, T., Chisholm, A., Harwatt, H., Horton, D., Jopson, A., Pooley, C., Schelderman, G. and Tight, M. 2009. Understanding Walking and Cycling: A Multi-Method Approach to Investigating Household Decision Making in Relation to Short Journeys in Urban Areas. Paper presented at *Cycling and Society Research Group Symposium*, University of Bolton, 7th September 2009. Available at: http://www.lec.lancs.ac.uk/research/society_and_environment/cycling/CSRG_Symposium_paper_Tim_Jones_Oxford_Brookes.pdf [accessed: 22 July 2013].

Marshall, S. 2005. *Streets and Patterns: The Structure of Urban Geometry.* Abingdon: Spon Press.

Mitra, R., Buliung, R. and Roorda, M. 2010. Built environment and school travel mode choice in Toronto, Canada. *Transportation Research Record*, 2156, 150–59.

Moudon, A.V. and Lee, C. 2003. Walking and bicycling: an evaluation of environmental audit instruments. *American Journal of Health Promotion: AJHP*, 18 (1), 21–37.

National Travel Survey 2011a. *Table NTS0308: Average number of trips by trip length and main mode: Great Britain, 2010.* [Online: UK Department for Transport]. Available at: http://www.dft.gov.uk/statistics/tables/NTS0308 [accessed: 27 July 2012].

National Travel Survey 2011b. *Table NTS0303: Average number of trips (trip rates) by main mode: Great Britain, 1995/97 to 2010.* [Online: UK Department for Transport]. Available at: http://www.dft.gov.uk/statistics/tables/nts0303/ [accessed: 27 July 2012].

Oliver, L.N., Schuurman, N. and Hall, A.W. 2007. Comparing circular and network buffers to examine the influence of land use on walking for leisure and errands. *International Journal of Health Geographics*, 6 (1), 41.

Ordnance Survey 2011. *OS MasterMap Address Layer and Address Layer 2 User guide (v1.4, 01/2011).* Southampton: Ordnance Survey.

Page, S., Cooper, A., Griew, P. and Jago, R. 2010. Independent mobility, perceptions of the built environment and children's participation in play, active travel and structured exercise and sport: the PEACH project. *International Journal of Behavioural Nutrition and Physical Activity*, 7.17, 1–10.

Pooley, C., Tight, M., Jones, T., Horton, D., Scheldeman, G., Jopson, A., Mullen, C., Strano, E. and Constantine, S. 2011. *Understanding Walking and Cycling: Summary of Key Findings and Recommendations.* Lancaster: Lancaster University.

Pooley, C., Jones, T., Tight, M., Horton, D., Scheldeman, G., Jopson, A., Mullen, C. and Strano, E. 2013a. *Promoting Walking and Cycling: New Perspectives on Sustainable Travel.* Bristol: The Policy Press.

Pooley, C., Horton, D., Scheldeman, G., Mullen, C., Jones, T., Tight, M., Jopson, A. and Chisholm, A. 2013b. Policies for promoting walking and cycling in England: a view from the street. *Transport Policy*, 27, 66–72.

Porta, S., Crucitti, P. and Latora, V. 2006. The network analysis of urban streets: a primal approach. *Environment and Planning B: Planning and Design*, 33 (5), 705–25.

Pucher, J. and Buehler, R. 2008. Making cycling irresistible: lessons from The Netherlands, Denmark and Germany. *Transport Reviews*, 28, 495–528.

Sabidussi, G. 1966. The centrality index of a graph. *Psychometrika*, 31, 581–603.

Saelens, B.E., Sallis, J.F. and Frank, L.D. 2003. Environmental correlates of walking and cycling: findings from the transportation, urban design, and planning literatures. *Annals of Behavioral Medicine: A Publication of the Society of Behavioral Medicine*, 25 (2), 80–91.

Scheurer, J. and Porta, S. 2006. *Centrality and connectivity in public transport networks and their significance for transport sustainability in cities.* Paper to World Planning Schools Congress, Global Planning Association Education Network, Mexico, 13–16 July 2006.

Van Dyck, D., Deforche, B., Cardon, G. and De Bourdeaudhuij, I. 2009. Neighbourhood walkability and its particular importance for adults with a preference for passive transport. *Health and Place*, 15 (2), 496–504.

Victoria Transport Policy Institute 2012. *Roadway Connectivity: Creating More Connected Roadway and Pathway Networks.* [Online: TDM Encyclopedia]. Available at: http://www.vtpi.org/tdm/tdm116.htm [accessed: 27 July 2012].

Walker, J.L. 2006. Opening up the black box: enriching behavioral models of spatial and travel choices. *Journal of Transport Geography*, 14, 396–8.

# Chapter 2
# Transport Rationales – The Link between Residential Location and Travel Behaviour

Petter Næss

## Introduction

Since the publication of the UN report 'Our Common Future' (World Commission on Environment and Development 1987), the issue of sustainable development has been a common challenge for all nations. Sustainable mobility is an important part of this and is commonly understood as mobility in accordance with the general principles of sustainable development. This involves, among other things, a volume of physical mobility, a modal-split and a transport technology that meet basic mobility needs in an efficient way, taking care of ecosystem integrity, limiting emissions to an environmentally sustainable level, and being safe and consistent with human health (Centre for Sustainable Transportation 2002, CIENS 2006). In line with this, limiting the growth of urban motoring has been an important policy goal in many European cities as well as in European sustainable mobility policies in general (OECD/ECMT 1994).

Urban land use influences the distances between homes, workplaces and service and leisure facilities and therefore the passenger base for public transport, thus potentially exerting a great influence on the travel behaviour of the inhabitants. Such relationships between urban spatial structures and travel make up an important part of the arguments in favour of the compact city as a sustainable urban form (CEC 1990, Jenks et al. 1996). Several studies on the relationships between urban land use and travel have shown associations between daily travel behaviour and the location of the dwelling. Suburbanites tend to travel longer overall distances and carry out a higher proportion of their travel by car than their inner-city counterparts, among whom walking and biking make up a higher share of the distance travelled. Such patterns have been found in a large number of cities in different corners of the world, including Paris (Mogridge 1985, Fouchier 1998), London (Mogridge 1985), New York and Melbourne (Newman and Kenworthy 1989), San Francisco (Schipper et al. 1994), Dutch urban regions (Schwanen et al. 2001), Austin (Zhou and Kockelman 2008), Athens (Milakis et al. 2008), Hangzhou (Næss 2009, 2010), Santiago de Chile (Zegras 2010), Copenhagen (Næss 2005, 2006, 2009, 2011), Oslo (Næss et al. 1995) and a number of other Nordic cities (Næss 2012).

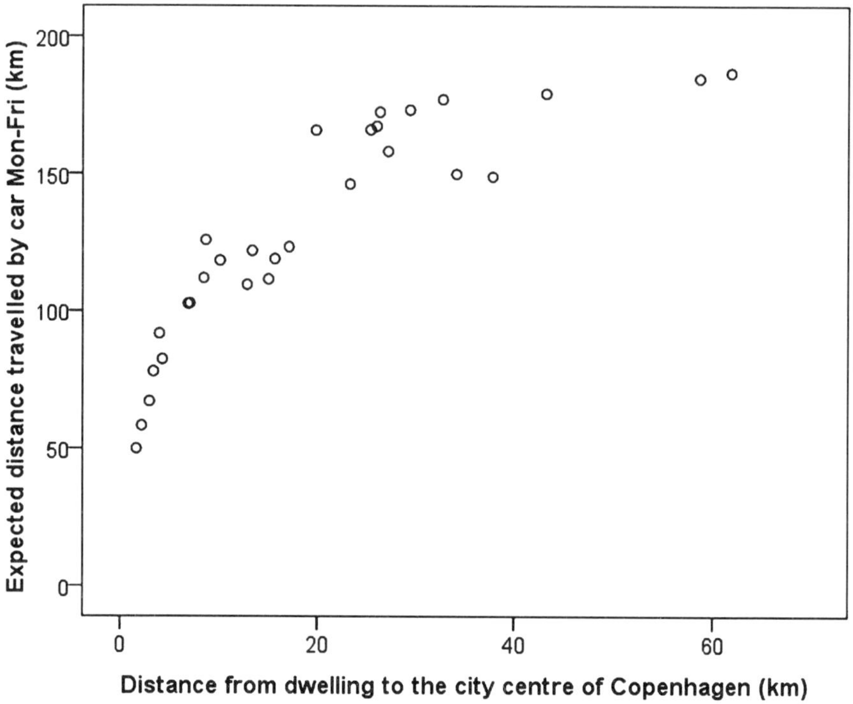


**Figure 2.1 Average expected travel distances by car (kilometre) over the five weekdays for each of the 29 investigated areas in Copenhagen Metropolitan Area. The circles are based on the respondents' actual values on each of the urban structural variables, and with socioeconomic variables, demographic variables and residential preferences kept constant at mean values. N = 1564 respondents, p = 0.0000.**

*Source:* Næss 2009.

This chapter will take a closer look at the role of transport rationales as links in the causal mechanisms by which residential location influences travel behaviour, using the metropolitan areas of Copenhagen and Hangzhou as case studies. In the above mentioned study of the Copenhagen Metropolitan Area, residents of the outer suburbs and exurbs tended to travel on average three to four times as long distances by car on weekdays as their inner-city counterparts (see Figure 2.1, based on data collected in 2001). Also, in the Chinese metropolitan area of Hangzhou, where car ownership is rapidly rising but was still substantially lower than in Copenhagen at the time of investigation (2005), a similar pattern was found. Residents living more than 10 kilometres away from the city centre tended to travel daily distances on average one and a half times longer than those living closest to the city centre (see Figure 2.2, to the left). Therefore, their likelihood of travelling by car or taxi was twice as high in comparison to their inner-city counterparts (see Figure 2.2, to the right). In both figures, controls have been made for differences in income, gender, household structure, transport-related residential preferences as well as a number of other demographic and socioeconomic variables (Næss 2009, 2010).

In both metropolitan areas, some influences on travel behaviour were also found in the distance from one's dwelling to the closest second-order and third-order centres. In the case of Copenhagen, this can be seen as drops and jumps in the expected mean travelling distances of the individual residential areas as compared to the general curvilinear tendency of increasing car travel with increasing distance from the main city centre (see Figure 2.1). In Hangzhou too, such tendencies could be identified for some travel behaviour variables, although they were not strong enough to show statistically significant effects on either total travelling distance or the likelihood to travel by car/taxi (Næss 2009).

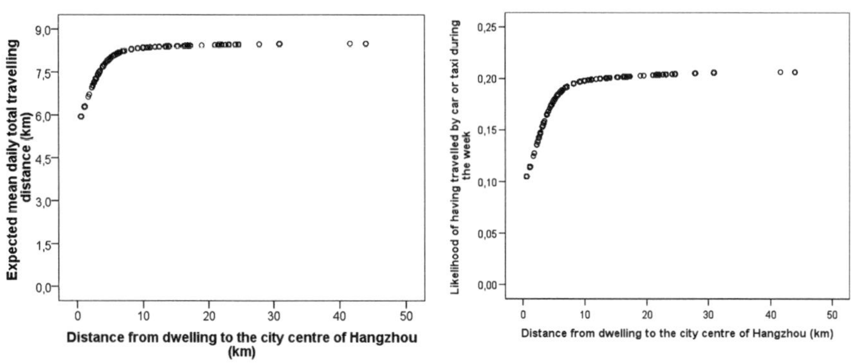


**Figure 2.2 Expected daily total travelling distance over the week (left) and likelihood of having travelled by car or taxi (right) among respondents living at different distances from the city centre of Hangzhou. N = 2091, p = 0.0000 for total travelling distance; N = 2115, p = 0.0003 for likelihood of having travelled by car or taxi.**

*Sources:* Næss 2010, 2007.

However, in order to substantiate that residential location is a (contributory) *cause* of such differences, we must show the basic mechanisms by which the location of dwellings influences travel behaviour. Examples illuminating the transport rationales of residents and how these rationales interact with spatial conditions in shaping routines for travel behaviour make up important elements in this endeavour. Transport rationales are here understood as the backgrounds, motivations and justifications that agents draw on when they make transport-relevant decisions about participating in activities, location of these activities, modes of transportation and the routes followed (Næss and Jensen 2005: 165).

The concept, which includes instrumental, safety-based, comfort-based, aesthetic as well as affective dimensions (Scheiner and Holz-Rau 2007, Steg et al. 2001, among others), has some overlap with the notion of 'mobility view' coined by Beckmann (2001). Combined with the spatial configuration of residences, employment and other facilities in a city or metropolitan area, it has been found in a number of different urban contexts that transport rationales produce some characteristic relationships between residential location and travel. The above-mentioned studies in the Copenhagen Metropolitan Area and Hangzhou Metropolitan Area have thrown light on such rationales. The mechanisms by which urban structure influences travel, identified in the qualitative interviews, are strongly consistent with the relationships between residential location and travel found in the quantitative parts of the two studies. These mechanisms can therefore help to explain the latter patterns.

This chapter seeks to show how various types of rationales for location of activities, travel modes, activity participation and route choice tend to strengthen or weaken aggregate-level relationships between residential location and travel. In particular, the influences of such rationales on the relative importance of residential location close to the main city centre and local centres, respectively, will be demonstrated. By investigating transport rationales in two very different (culturally, politically, economically and spatially) city regions, this chapter aims to illuminate whether there is some type of cross-contextual generality to the transport rationales and the causal mechanisms by which urban structure influences travel behaviour.

In the following two sections, the case cities and methods of the study will be presented. In the section on Rationales for Location of Activities these rationales will be dealt with in more length than other rationales, as they turn out to be of particular importance to the ways in which residential location influences travelling distances and thereby indirectly travel modes. Due to space constraints the section on Rationales for Mode Choice, Route Choice and Activity Participation will deal more briefly with these three rationales, especially the two latter ones, which will not be elaborated on in depth here. The Concluding Remarks will summarize the preceding sections as well as discuss ways in which relationships between residential location and travel are likely to develop in the future if the general mobility level continues to grow and if mobility is for some reason being constrained.

## Case cities

The Copenhagen Metropolitan Area has a population of about 1.8 million, is one of the largest urban areas of Northern Europe and acts as a major node for international air and rail transport. Its inner city has the unchallenged status as the region's dominant centre. The central structure of the Copenhagen Metropolitan Area can be characterized as hierarchic, with downtown Copenhagen as the main centre and the central parts of the five formerly independent outer-area towns, now engulfed by the major conurbation, as second-order centres. Alongside this are certain other concentrations of regionally oriented retail stores, and at a third level more local centre formations in connection with, among others, urban rail stations and smaller-size municipal centres.

Hangzhou is the capital of the Zhejiang province and is located 180 kilometres south-west of Shanghai in south-eastern China. Its metropolitan area includes 5 million inhabitants, about half of which live in the continuously built-up urban area of the city of Hangzhou. Similar to European cities, the historical urban cores of Chinese cities are usually the areas with the highest concentration of workplaces, retail stores and other service facilities. The Hangzhou Metropolitan Area is no exception. In addition to the major centre, the metropolitan area includes three second-order centres and six third-order centres. These centres, too, include a more or less comprehensive set of centre functions, but with a more narrow range and with a lower number of facilities within each category than in the main city centre.

Our two case cities differ considerably in terms of affluence level, cultural traditions and political conditions. In the Copenhagen Metropolitan Area, the mobility level has been high for decades, and three quarters of the respondents from our study belonged to households having one or more private cars at their disposal. In contrast, motor vehicle ownership has until recently been very low in Hangzhou, where sales of automobiles were restricted until 2004. Since then there has been an almost explosive growth in car ownership as well as ownership of other motor vehicles. Car ownership increased from 0.7 private cars per 100 households in 2001 to 15.4 cars per 100 households in 2007. At the time of our investigation (2005), 6 per cent of the respondents belonged to a household with a car. The respondents of the Copenhagen study travelled on average about eight times as long a distance on weekdays as the respondents of the Hangzhou study. In spite of these differences though, there are considerable similarities in the relationships between residential location and travel found in the two studies.

## Methods

The Copenhagen as well as the Hangzhou study were carried out by means of a combination of quantitative and qualitative research methods. The latter approach is still not very common, although some earlier examples of European studies exist (for example Røe 2001, Tillberg 2001, Næss and Jensen 2004, Scheiner 2005). Besides recording urban structural conditions from maps and aerial photographs, and visiting the investigated urban districts and residential areas, the investigation was based on a number of qualitative interviews (17 in Copenhagen and 28 in Hangzhou) and answers from individuals participating in questionnaire surveys (1,932 participants in Copenhagen and 3,155 participants in Hangzhou). More detailed information about the methods and results of the two studies is available in Næss (2005, 2006, 2009, 2010, 2013).

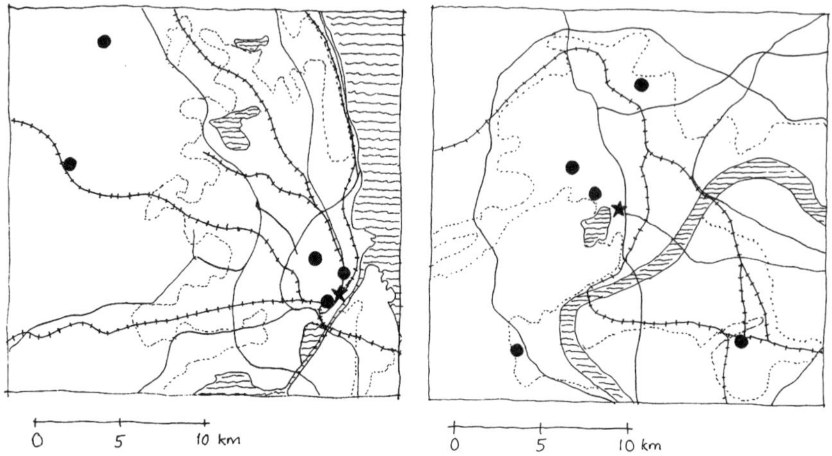


**Figure 2.3 Locations of the areas (indicated by filled circles) from where the interviewees of Copenhagen Metropolitan Area (left) and Hangzhou Metropolitan Area (right) were recruited. The metropolitan city centre of each urban region is marked with an asterisk. Demarcations of the continuous urbanized land in Copenhagen and Hangzhou are shown with dotted lines. Rail lines are shown as lines crossed by short perpendicular lines; main roads are shown as ordinary lines. Sea, river and lakes are indicated with parallel, horizontal wave-shaped lines.**

In this chapter, the primary source of evidence is the qualitative interview material. These interviews were semi-structured, based on interview guides and focused on the interviewees' reasons for choosing activities and corresponding locations, as well as travel modes, routes and the meaning attached to living in or visiting various parts of the city. The interviews were most often conducted in the homes of the interviewees and usually lasted between one and a half to two hours. In each of the city regions, the interviewees were recruited among the survey respondents from a selected number of the investigated residential areas. In the Copenhagen Metropolitan Area, nine interviewees were chosen from three inner-city areas and eight were recruited from two outer-suburban areas. One of the latter two areas was located close to an urban rail station whereas the other had very poor accessibility to public transport. In the Hangzhou Metropolitan Area, the interviewees were recruited from five of the investigated residential areas, representing typical inner-city neighborhoods (12 interviewees), suburban locations (11 interviewees) as well as one location close to the centre of one of the second-order towns (five interviewees). The interviews in Copenhagen were carried out by the author and a colleague from Aalborg University. In Hangzhou, the interviews were based on an interview guide translated into Chinese from its original English-language version and were carried out by a research team member from Zhejiang University. All interviews in both city regions were tape recorded and transcribed. The Chinese-language transcriptions were subsequently translated into English.

An interpretation scheme was developed in each case study as an important tool for the analysis. First, separate interpretations were written for each interview for each of our research questions. Thereupon, synthesizing interpretations were made, summarizing the information from all the interviews, which in total were 17 in Copenhagen and 28 in Hangzhou. These synthesizing interpretations were first made separately for each of the geographical locations from where the interviewees were selected, and thereafter for all interviewees together. By being required to make written interpretations of each interview in the light of each detailed research question, the interpretation required us to read and penetrate the transcribed interview texts in a far more thorough way than we would probably have done otherwise.

## Rationales for location of activities

Why does travel behaviour in the metropolitan areas of Copenhagen as well as Hangzhou depend more on metropolitan-scale than on local-scale built environment characteristics? The relative importance of metropolitan-scale and neighborhood-scale built environment characteristics to travel behaviour depends in particular on people's rationales for the location of the activities in which they participate.

In both city regions, the interviewees' location choice for their activities seems to be influenced by two main competing rationales, balanced against each other in different ways depending on a number of circumstances. These two rationales are:

- Choosing the best facility, and
- Minimizing the friction of distance

Each of these two rationales includes several more detailed aspects or sub-rationales. The rationale of choosing the best facility thus includes criteria related to the instrumental purpose of the activity, such as job content, salary, qualification requirements, and so on in the workplace and the diversity of commodities, prices, and so on in shops. To some extent though, the criteria also include cultural, symbolic or aesthetic properties of the locations, such as the 'atmosphere' of a particular place, as well as an aspect of variety-seeking. The rationale of minimizing the friction of distance (Lloyd and Dicken 1977) includes an aspect of minimizing the spatial (Euclidian) distance that must be travelled in order to reach the facility, minimizing the travelling time, minimizing the stress or physical efforts of travelling to the location (for example, in the form of changing between different means of transport) and minimizing the economic costs of the trip. Among our interviewees, especially in Hangzhou, the rationale of minimizing the friction of distance is often expressed in terms of choosing convenient locations. To a considerable extent, the sub-rationales under the rationale of minimizing the friction of distance overlap each other. But under certain conditions, such as congested roads, scarce parking or a particular configuration of the public transport lines, the fastest, least costly or most conveniently accessible locations may be different from the physically closest ones.

The friction of distance is a function of the time consumption, economic expenses and discomfort involved when travelling from one place to another. It is thus the inverse of the accessibility of the destination. Other things equal, the friction of distance will of course be at its lowest for the closest facilities. However, what is seen as the easiest accessible location varies with travel modes, depending on, among other things, the layout of the public transport network, the driving conditions along the road network and the conditions for walking and biking. For example, differences in parking conditions may imply that a somewhat more distant supermarket is more easily accessible by car than the closest supermarket. In other words, the friction of distance will be lower when driving to the former than to the latter location.

When talking about travel distance, the interviewees in both case studies were usually not very explicit regarding the ways in which distance matters. Some interviewees specifically referred to time consumption for travelling, but most often no particular reason was stated to explain why travel distance was important. Despite national differences, imperatives of not wasting either money or time are probably quite pervasive in the contemporary social and cultural contexts of both case regions, especially among the interviewees facing the strongest economic and time-geographical constraints. Arguably, the fact that so few interviewees explicitly mentioned money saving or time saving as an underlying motivation for minimizing travel distance most likely reflects the hegemonic status of these concerns, rendering it unnecessary to refer to them explicitly. Both of these rationales were also more often mentioned as concerns influencing choices of travel mode (see below).

Summarizing from above, the interviewees' activity locations appear to be influenced by two main rationales, each encompassing several sub-rationales:

1. Choosing the best facilities, including sub-rationales of:
   - Choosing facilities where the instrumental purpose of the activity can best be met;
   - Choosing facilities where social contacts can be maintained;
   - Choosing facilities matching the interviewees' cultural, aesthetic and symbolic preferences;
   - Variety-seeking.

2. Minimizing the friction of distance, including sub-rationales of:
   - Minimizing the spatial travelling distance;
   - Minimizing travel time;
   - Minimizing the stress or physical efforts of travelling to the destination;
   - Minimizing economic expenses associated with the trip.

Among our interviewees in both city regions, the sub-rationale of choosing facilities where the instrumental purpose of the activity can best be met is clearly more common than the sub-rationales associated with cultural, aesthetic and symbolic preferences and variety-seeking. The two latter sub-rationales exert some influence on the destinations of shopping and leisure trips among some interviewees. For example, a female office clerk living in one of Hangzhou's second-order centres sometimes joins her husband on trips to downtown Hangzhou, where he has an instrumental purpose (meeting or picking up someone), even though the interviewee herself has no special purpose. Her motivation for such a trip is based on an 'atmosphere' rationale or a mere wish for sightseeing. For 'bounded' activities like income-earning or studies, meeting the instrumental purpose of the activity is practically the only sub-rationale under the 'choosing the best facility' rationale.

Among the different travel purposes, trips to the workplace or a place of education along with visits to friends and relatives are where the longest distances are accepted. The destinations of visiting trips are defined entirely by the traveller's family relations and circle of acquaintances. Because the workplace or place of education is usually visited each weekday, while long visit trips are carried out far less frequently, journeys to work or education are the travel purposes accounting for the largest proportion of the travel distance on weekdays. The acceptable travel distance to work or education appears to increase in relation to how specialized the interviewees' work qualifications are, how many mobility resources they have at their disposal and how far away they live from the largest concentrations of work and educational opportunities.

For example, one interviewee in the Copenhagen Metropolitan Area, a 30-year-old computer engineer now living in a peripheral suburb, explained that he chose his present job without much consideration of the distance from their dwelling at that time:

> No, I just thought that I wanted employment, and then we would have to see where to go. And it did not matter where the workplace was located.

Similarly, a 38-year-old economist living in the same suburb said that he preferred to commute all the way to downtown Copenhagen instead of finding a less challenging job in a municipality closer to the residence:

> Surely, I would like something closer to home, but there are no such [relevant] jobs available to me here in the vicinity. Then it would have to be if I were interested in working in a municipal administration. … But that would be such

> a small workplace, and I simply want some more challenges. … Yes, for sure, most work opportunities for economists are in the city of Copenhagen.

For most of the interviewees in Copenhagen as well as Hangzhou, the choice of workplace appears to be influenced more by criteria such as job content, working conditions and salary than by the distance (in travel time or kilometres) from the dwelling. The interviewees usually admitted that there is a balance between the two criteria, but they generally seemed to be willing to travel quite far, if necessary, in order to find a job matching their qualifications, interests and expectations regarding salary level. The following quote from a Hangzhou interviewee, a male office clerk living in the semi-central district of Cuiyuan, is illustrative:

> I think if your job can satisfy both your personal value and salary, which will be great. I think the distance can be overcome if your job can meet these points.

This willingness does not seem to be influenced by the location of the residence, but rather to some extent by gender and family responsibilities (some women with small children say that they do not want to work far away from home as long as their children are small). The importance attached to different qualities of the workplace seems to vary somewhat between population groups, with the highest emphasis on job content among academics with a specialized education. But besides being predisposed by their disciplinary specialization to seek quite narrow niches in terms of job content, academics also face a job market where salary differences between the relevant jobs are modest, as emphasized by one of the Hangzhou interviewees.

For shops, the range and quality of commodities, the price level, the service level and the friendliness of employees and the degree of crowding are mentioned by interviewees as aspects influencing the attractiveness of stores. In addition, some Hangzhou interviewees who have a car at their disposal mention parking conditions as an aspect influencing which facility is considered to be the best. This latter criterion overlaps to a considerable degree with the rationale of minimizing the friction of distance, as shortage of parking places implies that a store further away from the destination has to be chosen, or else the destination must be reached by a slower mode.

In daily travel, some trips are more fixed and basic than others. Often, such a trip makes up the stock of a trip chain. Other travel purposes are then 'hitched' on to this stock trip. For example, by choosing a well-assorted store along the route followed anyway on the way home from work, the rationale of distance limitation can be combined with the rationale of choosing the best facility. In this way, the longer distances to shops, typical for residences in the outskirts of the city, can to some extent be compensated. This kind of adaptation is very common among our suburban interviewees, especially in the Copenhagen region.

When it comes to leisure trips, the choice among facility categories depends strongly on the interests and lifestyle of the person in question. Despite this though, quality differences within each facility category matter as well. For example, a distant, but larger and more beautiful forest may be preferred for outings rather than a local forest. Variety-seeking is particularly important as a rationale for choosing locations for outdoor recreation on weekends or holidays, where much of the motivation is the aesthetic experience of new landscapes. Yet, one single natural area – at least if the area is of a certain size – may also offer opportunities for numerous trips where new aspects of the landscape are discovered each time. Variety-seeking as a motive for leisure activities was mentioned explicitly only among some Hangzhou interviewees and was not identified in the Copenhagen Metropolitan Area study. It has, however, been discussed in the literature on leisure travel (Stauffacher et al. 2005).

The sub-rationale of choosing facilities where social contacts can be maintained was also only encountered in Hangzhou. This sub-rationale appears to also be important for certain leisure activities such as visits to teahouses, restaurants and to some extent outdoor recreation areas, as illustrated by this quote from a female office clerk living in the suburban district of Banshan:

> Yes, we meet in downtown teahouse. I choose the place according to convenience for all of us.

For some other activities, the social contact in question is mandatory or necessary for the activity. Using the terminology of Hägerstrand (1970), it should be understood as a 'coupling constraint' rather than a rationale for activity location. Examples of activities here include attending the workplace or school, or participating in tennis games or football matches. However, for several other non-work activities such as shopping, swimming and hill climbing, the possibility of social contact in connection with the activity appears to play little or no role for the choice of location.

In our research, we have also found that a young interviewee visits certain locations (for example, the downtown areas of Hangzhou and one of the second-order towns) with 'adventure-seeking' as the main purpose. The possibility of experiencing 'adventure' (for example, in the form of random social contacts) is arguably a part of the particular opportunities of downtown centres, where many people pass by on their way to and from the workplace, leisure facilities and so on. 'Adventure-seeking' could thus be considered a second-order sub-rationale under the sub-rationale of social contact.

Table 2.1 summarizes how the different rationales contribute to the influences on travelling distances from the location of residences relative to the main concentration of facilities and to local facilities, respectively. The different rationales have all been identified in the qualitative interviews.

**Table 2.1 Contributions of various rationales for location of activities to the relationships between residential location and travel**

| Rationales for activity location | Frequency of occurrence | Influence on activity location | Influence on the relationship between the amount of travel and the distance from the dwelling to the main centre of the metropolitan area | Influence on the relationship between the amount of travel and the distance from the dwelling to local facilities |
|---|---|---|---|---|
| Choosing facilities where the instrumental purpose of the activities can best be met | Emphasized by nearly all interviewees, but its importance varies between activity types and between individuals. | Tends to make the interviewees consider a large number of facilities within each facility category as potential locations of their activities, regardless of the distance from the dwelling to these facilities (as long as some quite wide threshold distance is not exceeded). | Contributes strongly to this relationship by increasing the likelihood of travelling to the large concentration of facilities in the inner parts of the metropolitan area, but also because of downtown's role as an approximate point of gravity for all peripheral destinations. | Contributes to a certain weakening of this relationship by increasing the likelihood of choosing distant facilities rather than local ones. |
| Choosing facilities where social contacts can be maintained | Emphasized by several Hangzhou interviewees as a criterion for choosing which teahouses, restaurants, and so on to visit. Not mentioned among Copenhagen interviewees. | Tends to make interviewees choose facilities not only based on their own preferences, but on the common preferences (in terms of accessibility, quality criteria, and so on) of a group of friends. | Contributes somewhat to strengthen this relationship because of downtown's role as an approximate point of gravity for the housing stock and its high accessibility by public transport. | May contribute somewhat to strengthen this relationship insofar as the groups of friends who decide to meet at teahouses and so on live in the same local district. |

| | | | | |
|---|---|---|---|---|
| Choosing facilities matching the interviewees' cultural, aesthetic and symbolic preferences | Emphasized by several interviewees as a criterion for 'non-bounded' trips, particularly in connection with location of leisure activities and also sometimes shopping. | Tends to make interviewees choose certain picturesque, reputable or historically interesting areas as locations for leisure and shopping activities. | Contributes somewhat to strengthen this relationship because several of the culturally, aesthetically and symbolically most attractive areas are either located close to the downtown area or at locations more easily accessible from the inner city than from most of the outer parts of the metropolitan area. | Contributes to a certain weakening of this relationship by making respondents bypass facilities in local centres where there is less 'urban atmosphere' and historically interesting qualities than in the downtown area. |
| Variety-seeking | Mentioned or indicated by some Hangzhou interviewees as a reason for shifting between different recreational areas or supermarkets. | Combined with rationales of choosing the best facility, variety-seeking tends to make interviewees sometimes choose more distant facilities than the closest one matching the interviewee's quality criteria. | Since a large number of alternative facilities can usually be found close to the dwellings of inner-city residents, variety-seeking is not likely to imply significantly increasing travelling distances among these residents. Due to the lower density of facilities in the outer parts of the metropolitan area, the variety-seeking of outer-area residents is more likely to imply increased travelling distances. The variety-seeking rationale thus probably contributes to a slight strengthening of the relationship between the amount of non-work travel and the distance from the dwelling to the main centre of the metropolitan area. | By making interviewees sometimes choose more distant locations than they would otherwise have done, variety-seeking tends to reduce the use of local facilities and thus tends to weaken the relationship between the amount of non-work travel and the distance from the dwelling to the closest local centre. |

| Rationales for activity location | Frequency of occurrence | Influence on activity location | Influence on the relationship between the amount of travel and the distance from the dwelling to the main centre of the metropolitan area | Influence on the relationship between the amount of travel and the distance from the dwelling to local facilities |
|---|---|---|---|---|
| Minimizing the spatial travelling distance | Emphasized by nearly all interviewees, in particular those without a car. Thresholds for acceptable distances vary between activity types and between individuals. | Tends to make the interviewees limit their choices of facilities for a given type of activity to those facilities which are accessible within a certain geographical radius, and to choose the closest facility meeting his/her quality criteria. Threshold distances are usually widest for workplaces and shortest for daily necessity shopping. | Contributes to some extent to this relationship, both because the facilities in the central districts of each city are the closest opportunities for inner-city residents, and because of the shortage of facilities in the periphery. | Contributes strongly to this relationship by increasing the likelihood of choosing local facilities rather than more distant ones. |
| Minimizing travel time | Although mentioned explicitly by relatively few interviewees, time saving is probably of quite general importance as a sub-rationale contributing (together with distance minimizing) to minimizing the friction of distance. Thresholds for acceptable time consumption vary between activity types and between individuals. | Tends to make the interviewees limit their choices among facilities for a given type of activity to those facilities which are accessible within a certain travel time, and to choose the facility meeting his/her quality criteria which can be reached with the least time consumption. Thresholds for travel time are usually widest for workplaces and shortest for daily necessity shopping. | May induce some car drivers to choose, for example, large suburban supermarkets instead of central-city shops. Contributes nevertheless to some extent to the relationship between the distance from the residence to downtown and the amount of travel, due to the function of the urban centre as geographical point of gravity. | Contributes to this relationship because it will usually take a short time to go to local facilities. But because travel speeds by car will often be higher when going to, for example, a more distant shopping mall with ample parking space, the influence of this rationale is not as strong as the influence of the rationale of limiting geographical distances. |

| | | | | |
|---|---|---|---|---|
| Minimizing the stress or physical efforts of travelling to the destination | Emphasized in particular among Hangzhou interviewees who do not have any private motorized vehicle at their disposal. | Tends to make interviewees travelling by non-motorized modes limit their travelling distances, and to make interviewees travelling by public transport avoid destinations necessitating several and/or cumbersome shifts between different public transport lines. | Tends to weaken this relationship somewhat by increasing the propensity of suburbanites without a car at their disposal – in particular those living in areas with poor public transport services – to limit their choices among facilities to those available locally. | Contributes to this relationship by increasing the likelihood of choosing local facilities rather than more distant ones. |
| Minimizing economic expenses associated with the trip | Mentioned explicitly by only a few interviewees, but it is hard to imagine that this does not play some role as a sub-rationale contributing to minimizing the friction of distance. | Tends to make interviewees use facilities within walking or biking distance to a higher extent than what they would otherwise do, and to choose destinations for car trips where it is not necessary to pay high parking fees. Contributes also somewhat to a general limitation of travelling distances by motorized modes. | May induce some car drivers to choose, for example, suburban stores and leisure facilities instead of downtown facilities. Contributes nevertheless to some extent to this relationship, both because the facilities in the central districts of each city are the closest opportunities for inner-city residents, and because of the shortage of facilities in the periphery. | Contributes clearly to this relationship because local facilities will usually be the ones that can be reached with the smallest economic expenses. |

Among the rationales influencing the interviewee's location of activities, the relationship between the amount of transport and the distance from the residence to the main centre of the urban region tends to be particularly strengthened by the rationale of choosing facilities where the instrumental purpose of activities can best be met. Additionally though, this decision is influenced by the rationales of social contacts and of cultural, aesthetic and symbolic preferences as well as to a lesser extent the rationales of variety-seeking, minimizing spatial travelling distance, minimizing travel time and minimizing economic expenses. The only rationale among those identified that contributes to somewhat weaken this relationship is the rationale of minimizing the stress or physical efforts of travelling.

The relationship between the amount of transport and the distance from the residence to the closest local centre tends to be strengthened in particular by the rationale of minimizing spatial travelling distance, but also by the rationales of social contacts, minimizing travel time, minimizing the stress or physical efforts of travelling and minimizing economic expenses. This relationship seems to be weakened by rationales of choosing facilities where the instrumental purpose of the activities can best be met; cultural, aesthetic and symbolic preferences; and variety-seeking.

It might perhaps seem paradoxical that the rationale of travel time minimizing contributes in the same direction both for relationship to main centre and to local facilities. As illustrated in Table 2.1, emphasis on limitation of time consumption may induce some car drivers to choose, for example, suburban shopping malls instead of shops located in congested inner-city areas. Nevertheless, the time saving rationale contributes to some extent to the relationship between the distance from the residence to downtown and the amount of travel. Partly, this is due to the function of the urban centre as geographical point of gravity. Moreover, due to a denser population as well as the presence of jobs and other facilities, inner-city facilities will also be considered the closest ones for a considerable number of inhabitants. The travel time saving rationale contributes to the relationship between the amount of travel and the distance from the dwelling to local facilities too though, since it usually takes a short time to go to local facilities.

Our research shows that the propensity for using local facilities depends partly on which facilities exist in the proximity of the dwelling and partly on the competition from non-local facilities. In the districts next to the downtown area, a relatively broad supply of local facilities often exists, but at the same time there is a strong competition from facilities in the city centre. Conversely, the local supply of facilities is often more modest in the outer parts of the metropolitan area, but the long distance to the concentration of facilities found in the downtown area at the same time weakens the competition from the latter facilities.

**Rationales for modal choice, route choice and activity participation**

The interviewees' *choices of travel modes* are influenced by a number of different and interconnected rationales. These rationales have been classified into two main groups:

- Rationales concerning the efficiency of the movement from origin to destination
- Rationales concerning the process of moving from origin to destination

The first of these two groups includes concerns related to the time consumption, economic costs and accessibility benefits of travelling by different modes. The second group includes concerns related to physically, psychologically and socially positive or negative aspects associated with travelling by a particular mode.

Among the rationales concerning the efficiency of the movement from origin to destination, the following appear to be the most important to our interviewees:

- Time saving
- Flexibility
- Expansion of the radius of action
- Money saving

The rationales concerning physical, psychological and social aspects associated with the process of travelling include:

- Comfort
- Limitation of physical efforts
- Relaxation
- Safety
- Aversion to frustration
- Physical exercise
- Enjoyment of surrounding environment
- Affective dislike or preference for a particular mode
- Lifestyle signalling
- Habits
- Demonstration of wealth and status

The rationale of time saving generally leads interviewees to choose those modes of travel that can bring them as fast as possible from their origin to their destination. Among interviewees who do not have a car at their disposal, this implies a preference for the bike on short trips, public transport for long trips, and avoiding the rush hours when travelling by bus. Among car owners, the time saving rationale encourages choosing a car for long trips and non-motorized modes for short trips when driving (including the walking time to and from parking) would be more time-consuming than a non-motorized mode. The time saving rationale has a stronger influence on the choice of travel mode in proximity to the main city centre, since more slowly-moving traffic in inner-city areas makes up an incitement for residents of these areas to choose non-motorized modes or public transport.

Since long trips are very time-consuming as well as physically exhausting if they are made by non-motorized modes (in particular by foot), rationales of time saving and limitation of physical efforts will logically imply a dependence on travel modes in relation to trip distances. Trip distance thus appears to have the role of an intermediate rationale through which residential locations with shorter travelling distances are also conducive to a higher share of non-motorized travel.

The rationale of flexibility generally leads to a preference for individual modes of travel rather than public transport (due to its often rigid layout of lines and time schedules). For short and medium-long trips, this implies a preference for the bike (or walking for shorter distances) rather than bus. For longer trips, the flexibility rationale leads to a preference for the car, motorcycle or e-bike. This flexibility rationale may strengthen the influence of proximity to the main city centre on modal choice, since public transport is less flexible in the periphery and car traffic more hampered in the inner city.

The rationale of expanding the radius of action is related to the rationale for choosing the best facility described in the previous section, as the use of motorized modes, in particular the car, expands the geographical area within which relevant facilities can be chosen. For a particular trip, such as one with the purpose of shopping, a car may be chosen in order to visit a broader range of shops than would otherwise be possible within an acceptable level of time consumption. The radius of action rationale strengthens the relationship between proximity to the city centre and modal choice, since the incentive to increase the radius of action is higher in outer areas where a more narrow range of facilities is available within a short distance from the dwelling. To some extent, this rationale also contributes to more car trips when living far away from the closest local centre.

The rationale of money saving strengthens the influence of a central residential location (especially proximity to the main city centre) on modal choice, since gasoline consumption per kilometre is higher and parking more expensive in inner-city areas. Additionally, the money saving rationale may act as a catalyzer, inducing residents to realize the potential for non-motorized travel when living close to relevant destinations. The latter also applies to the rationales of physical exercise, experience of surroundings and aversion to frustration, which may contribute to realizing a potential for walking and cycling when distances to facilities are moderate.

Although the remaining rationales for modal choice are also interesting, they appear to be less important for understanding the ways in which residential location influences travel behaviour. Therefore, we will not elaborate on these rationales.

Our interviews indicate that people's *activity patterns* are to some extent adapted to the availability of facilities in the proximity of the dwelling. The interviewees still rarely give up activities completely as a result of moving to a different urban structural situation. According to our survey data, 'distance decay' in the form of reduced activity participation when living far away from relevant facilities is not very pronounced among our respondents. In particular, this applies to workforce participation. We find no evidence of reduced employment frequencies among people living far away from the main concentrations of workplaces, and only very modestly a tendency of 'teleworking' among respondents with long commuting distances. For leisure activities, the total amount of participation does not vary much with residential location, although the composition of activities does to some extent, depending on where the relevant facilities can be found. In general, the relationships between residential location and the frequencies of activity participation are relatively weak, and distance decay therefore only to a small extent counteracts the overall tendency of increased travelling distances and more frequent car trips when living far away from the main concentrations of facilities.

The relationships between the amount of travel and the distance from the dwelling to the city centre and local centres are theoretically based on the assumption that there is no systematic difference between inner-city and outer-area residents in the extent of any deviations from the shortest route. The rationales for *route choice* generally support this assumption. Apart from the rationale of physical exercise, which may lead to considerable deviations from the shortest route when making exercise trips by foot or by bike, the interviewees are not apt to make long detours from the shortest route to daily destinations. Since the need for extending the length of non-motorized trips in order to obtain sufficient exercise is higher if the destination is located closer rather than far away, the tendency to make 'exercise-motivated detours' is likely to be more widespread among inner-city dwellers than among their outer-area counterparts. However, from an environmental perspective, this compensatory mechanism – which contributes to the somewhat increased travel distance among exercise-minded inner-city residents – is unimportant, since these non-motorized trips don't consume any fossil energy and don't generate greenhouse gas emissions.

## Concluding remarks

The rationales for location of activities, choice of transport modes, activity participation and route choice identified in qualitative research are important links within the mechanisms by which urban structures influence travel behaviour. Notably, inhabitants show a tendency in modern cities to emphasize (within some threshold distances) the possibility of choosing among facilities rather than proximity, meaning that the amount of travel is influenced to a higher extent by the location of the residence in relation to concentrations of facilities than the distance to the closest single facility within a category. Daily travelling distances therefore tend to be more influenced by the distance from the dwelling to the city's main concentration of facilities (usually the inner city) than by its distance to local centres.

Internationally, relatively few studies have included qualitative interviews in order to identify such rationales, reflecting a general dominance of quantitative research within the field of land use and transport studies. Among the few qualitative investigations of transport rationales apart from the ones presented in this study, a high proportion has been carried out in the Nordic context (Røe 2001, Tillberg 2001, Nielsen 2002, Næss and Jensen 2002, 2004). These studies have taken place in different urban contexts, including the Norwegian capital of Oslo, medium-sized towns such as Aalborg (Denmark) and Gävle (Sweden), and the small town of Frederikshavn in Denmark. Nevertheless, the rationales identified in the different urban regions and the ways in which they mediate the influence of urban structure on travel are highly consistent. In Berlin, with its unique history, the cultural legacy of the divided city was still found to generate travel patterns differing considerably from what could be expected from the physical and spatial structure of the city (Scheiner 2005).

More typically, according to Røe's interviews with 15 individuals living at different locations in Oslo, inner-city living allows highly mobile and spatially flexible individuals opportunities to choose between a large number of services in the vicinity of the dwelling as well as elsewhere in the urban region. On the other hand, people belonging to the same mobility lifestyle group and living in the outskirts of the city need to travel long distances in order to meet their preferences. For inner-city residents with less flexible mobility lifestyles, the availability of a wide range and number of facilities within a relatively short distance from the dwelling allows for a wide variety of opportunities despite these interviewees' relatively confined geographical radius of action. Members of lower mobility lifestyle groups in the suburbs, however, experience several spatial constraints and a lower degree of freedom of choice, especially if the area is poorly served by public transportation (Røe 2001: 221).

The similarity of transport rationales across the widely differing contexts of the Copenhagen and Hangzhou metropolitan areas suggests that there may be a high degree of generality in the basic mechanisms through which urban form influences travel behaviour. The prioritization between the rationales differs between different population groups, depending on their mobility resources, professional qualifications, lifestyles and values as well as the opportunities provided by the spatial configuration of facilities and the transport system. In Western cities, and increasingly also in relatively affluent Asian urban contexts like the Hangzhou Metropolitan Area, most people tend to prioritize opportunities for choice above geographical proximity when choosing locations for their activities, especially regarding jobs and out-of-home leisure activities.

If the general level of mobility in society continues to increase, distance minimizing will become even less important, and the rationale of choosing the best facility will additionally increase its salience. Travel behaviour will then be increasingly influenced by the location of one's dwelling relative to the main concentrations of facilities, while proximity to local centres will become less important. Densification close to these concentrations (usually in the main city centre and its surroundings) will then be most in line with goals of reducing urban motoring and hence be favourable from a sustainable mobility perspective. If, however, the level of mobility is reduced in the future, such as due to environmental policy instruments or higher fuel prices, proximity to more local centres may again become important. Such reduction in general levels of mobility is arguably required if the frequently recommended polycentric models of urban development are to be compatible with sustainability criteria. This fits well with the analysis of Brotchie (1984), who argues that a decentralized settlement structure is the most energy efficient and requires the least transport if the level of physical mobility in the society is low. In such a situation, the distance decay will be high, with rationales of distance minimizing outweighing those of choosing the best facility. In a high-mobile society though, the deterrent of distance will be low, with rationales of choosing the best facility generally dominating any rationale to minimize distance (considerate of some threshold of acceptable travel time). If a peripheral settlement is to function in a self-contained way in a high-mobility society then, it will need to be located outside the catchment area of competing centres.

## Acknowledgements

The author would like to thank the Zhejiang University members of the research team: Professor Wenyao Yin, Weihong Ma, Yinmei Yao, Fen Li and especially Hui Yan, who carried out the interviews in Chinese and transcribed them into English. The research was funded by a grant from Volvo Research and Educational Foundation.

## References

Beckmann, J. 2001. *Risky Mobility. The Filtering of Automobility's Unintended Consequences.* Copenhagen: University of Copenhagen.

Brotchie, J.F. 1984. Technological change and urban form. *Environment and Planning A*, 16, 583–96.

CEC 1990. *Green Paper on the Urban Environment.* Luxembourg: Office for the Official Publications of the European Communities.

Centre for Sustainable Transportation 2002. Definition and vision of sustainable transportation. Available at: http://cst.uwinnipeg.ca/documents/Definition_Vision_E.pdf [accessed: 07 November 2012].

CIENS 2006. *Strategies and Actions for Common Research.* Oslo: Oslo Centre for Interdisciplinary Environmental and Social Research.

Fouchier, V. 1998. *Urban density and mobility in Ile-de France region.* Paper presented at The UN-ECE 8th Conference on Urban and Regional Research, Madrid, 8–11 June 1998.

Hägerstrand, T. 1970. What about People in Regional Science? *Papers of the Regional Science Association*, 24, 7–21.

Jenks, M., Burton, E. and Williams, K. (eds) 1996. *The Compact City: A Sustainable Urban Form*? London: E & FN Spon.

Lloyd, P.E. and Dicken, P. 1977. *Location in Space – a Theoretical Approach to Economic Geography.* London: Harper & Row.

Milakis, D., Vlastos, T. and Barbopoplos, N. 2008. Relationships between urban form and travel behaviour in Athens, Greece. A comparison with Western European and North American results. *European Journal of Transport Infrastructure Research*, 8, 201–15.

Mogridge, M.H.J. 1985. Transport, land use and energy interaction. *Urban Studies*, 22, 481–92.

Newman, P.W.G. and Kenworthy, J.R. 1989. *Cities and Automobile Dependence.* Aldershot: Gower Publications.

Nielsen, T.S. 2002. *Boliglokalisering og transport i Aalborg (Residential Location and Transport in Aalborg).* Aalborg: Aalborg University.

Næss, P. 2005. Residential location affects travel behavior – but how and why? The case of Copenhagen Metropolitan Area. *Progress in Planning*, 63 (2), 167–257.

Næss, P. 2006. *Urban Structure Matters.* London/New York: Routledge.

Næss, P. 2007. *Residential Location and Travel in Hangzhou Metropolitan Area*, NIBR report 2007: 1. Oslo: Norwegian Institute for Urban and Regional Research.

Næss, P. 2009. Residential self-selection and appropriate control variables in land use-travel studies. *Transport Reviews*, 29, 293–324.

Næss, P. 2010. Residential location, travel and energy use: the case of Hangzhou Metropolitan Area. *Journal of Transport and Land Use*, 3, 27–59.

Næss, P. 2013. Residential location, transport rationales and daily-life travel behavior: the case of Hangzhou Metropolitan Area, China. Forthcoming in: *Progress in Planning*, 79 (1).

Næss, P. and Jensen, O.B. 2002. Urban land use, mobility and theory of science: exploring the potential for critical realism in empirical research. *Journal of Environmental Policy and Planning*, 4 (4), 295–311.

Næss, P. and Jensen, O.B. 2004. Urban structure matters, even in a small town. *Journal of Environmental Planning and Management*, 47, 35–56.

Næss, P. and Jensen, O.B. 2005. *Bilringene og cykelnavet. Boliglokalisering, bilafhængighed og transportadfærd i Hovedstadsområdet (The Car Tires and Bike Hub. The Residential Location, Car Dependency and Travel Behavior in Copenhagen Metropolitan Area).* Aalborg: Aalborg University Press.

OECD/ECMT 1994. *Final report of the joint OECD/ECMT project group on urban travel and sustainable development*. Paris: OECD/European Conference of Ministers of Transport.

Røe, P.G. 2001. *Storbymenneskets hverdagsreiser. Sammenhenger mellom bosted, livsstil og hverdagsreisepraksis i et senmodern perspektiv (The Everyday Travel of City Dwellers. A Late-Modern Perspective on the Relationships between Place of Dwelling, Life Style and Everyday Travel Practice)*. Trondheim: Norwegian University of Science and Technology.

Scheiner, J. 2005. Daily mobility in Berlin: on 'inner unity' and the explanation of travel behaviour. *European Journal of Transport and Infrastructure Research*, 5 (3), 159–86.

Scheiner, J. and Holz-Rau, C. 2007. Travel mode choice: affected by objective or subjective determinants? *Transportation*, 34, 487–511.

Schipper, L., Deakin, E. and Spearling, D. 1994. *Sustainable Transportation. The Future of the Automobile in an Environmentally Constrained World*. Paper presented at a research seminar arranged by the Board of Communications Research, Stockholm, September 23, 1994.

Schwanen, T., Dieleman, F.M. and Diest, M. 2001. Travel behaviour in Dutch monocentric and policentric urban systems. *Journal of Transport Geography*, 9, 173–86.

Stauffacher, M., Schlich, R., Axhausen, K.W. and Scholz, R. 2005. The diversity of travel behaviour: motives and social interactions in leisure time activities. *Arbeitsberichte Verkehr- und Raumplanung*, 30x, IVT, ETH Zürich, Zürich. Available at: http://fex-dev2.ethz.ch/eserv/eth:28934/eth-28934-01.pdf [accessed: 07 November 2012].

Steg, L., Vlek, C. and Slootegraaf, G. 2001. Instrumental-reasoned and symbolic-affective motives for using a motor car. *Transportation Research Part F*, 4, 151–69.

Tillberg, K. 2001. *Barnfamiljers dagliga fritidsresor I bilsamhället – ett tidspussel med geografiska og könsmässiga variationer (Daily Leisure Trips of Families with Children in a Car-based Society – a Time Puzzle Involving Geographical and Gender Difference)*. Geografiska regionstudier nr. 43. Uppsala: Uppsala University.

World Commission on Environment and Development 1987. *Our Common Future*. Oxford/New York: Oxford University Press.

Zhou, B. and Kockelman, K. 2008. Self-selection in home choice: use of treatment effects in evaluating the relationship between the built environment and travel behavior. *Transportation Research Record*, 2077, 54–61.

## Chapter 3
# Spatial (Mis)match and its Consequences for Mode Choices

Veronique Van Acker, Ben Derudder and Frank Witlox

### Introduction

Many studies have analysed the relationship between the built environment and mode choice, but underlying behavioural mechanisms remain somewhat less well understood. Higher density levels, more diversity and better local accessibility are often believed to result in less car use and more public transport, cycling and walking (for a more comprehensive review, see, for example, Handy 2002, 2005, van Wee 2002, Van Acker and Witlox 2005, Bartholomew and Ewing 2009). However, not all people residing in high-density, diverse and accessible neighbourhoods travel by default using public transport, biking or walking instead of using their cars. This is (partly) due to differences in more subjective and behavioural influences such as perceptions (Van Acker et al. 2010). It might be possible that one person perceives the built environment as unsafe, preventing him or her from walking, whereas another person might feel relatively safe walking. Recently, attempts have been made to incorporate such subjective influences into land use-travel behaviour interaction models (for example, Kitamura et al. 1997, Bagley and Mokhtarian 2002, van Wee et al. 2002, Scheiner and Holz-Rau 2007). However, only a few of these studies have questioned whether these subjective influences correspond to the objective reality. For example, a neighbourhood is objectively evaluated as pedestrian friendly (for example, low motorized traffic levels, availability of sidewalks), but an individual with a specific lifestyle might still consider this neighbourhood as unsafe. Therefore, it would be interesting to balance objective variables against more subjective variables.

One exception to this is the series of studies done by Schwanen and Moktharian (2005a, b), which focuses on the concept of residential neighbourhood type dissonance, seen as the mismatch between preferred and actual type of residential location. They found that travel behaviour of the mismatched individuals corresponds to the matched residents of the actual neighbourhoods, suggesting that the influence of the built environment remains important despite mismatched spatial preferences. However, it might be interesting to also know how people perceive their current residence and how this corresponds with the objectively measured spatial characteristics of their residential neighbourhood. This would offer insights in the accuracy of someone's spatial knowledge about their actual residential neighbourhood. For example, the distance between one's residence and the nearest bus stop can be objectively measured, but there are no guarantees that a short distance is also perceived as such. Non-public transport users might especially be unaware that a bus stop is within close distance of their residence. In this chapter, we will focus on the travel consequences of such (mis)matched spatial perceptions. It is important to know whether travel decisions are influenced by the design and layout of the residential neighbourhood (for example, pedestrian friendly neighbourhoods that actually encourage walking) or whether personal influences such as perceptions are more important. The issue of spatial (mis)match thus questions the sustainability of spatial planning policies aimed at, for example, the densification and mixing of land use.

The analysis reported in this chapter is based on data from an Internet survey on lifestyles and leisure mobility in Flanders (Belgium) that additionally questioned the respondents' perceptions of their current residential neighbourhood. By adding spatial information from other land use databases, spatial perceptions can be balanced against the objective spatial characteristics of the respondents' current residential neighbourhood. As a result, the consequences of the (mis)matched spatial perceptions on mode choices for leisure activities will be evaluated.

## Data

Current travel surveys generally lack information on subjective influences such as perceptions. Therefore, we conducted an Internet survey between May 2007 and October 2007. The survey was made known to students and staff members of the University of Antwerp and the Faculty of Sciences at Ghent University, and an announcement was published in regional information magazines of several villages in the larger urban region of Ghent (Flanders, Belgium). In total, 2,363 people completed the survey, of which 1,626 were retained after data cleaning for further analysis. Figure 3.1 illustrates the residential locations of these respondents.

Despite our efforts, we did not obtain a well-balanced sample. Women, married couples, people with full-time employment and younger people are overrepresented. But the most remarkable difference is in education. Highly-educated respondents are heavily overrepresented in the sample; 66 per cent have a college or university degree, which is considerably higher than the average of 25 per cent for Flanders. Although the sample is not representative of the entire population of Flanders, we feel that this does not devalue it for our research purposes and results. Our purpose is to model relationships among variables, not to ascertain the univariate distributions of variables in isolation. Our analysis can still properly capture the conditional influence of having a given level of education on travel behaviour, even if the proportion of people having that amount of education differs between our sample and the population. The sample also permits demonstration of our premise that, conditional to a given level of education, subjective variables such as personal perceptions can still explain a significantly additional amount of variance in mode choices.

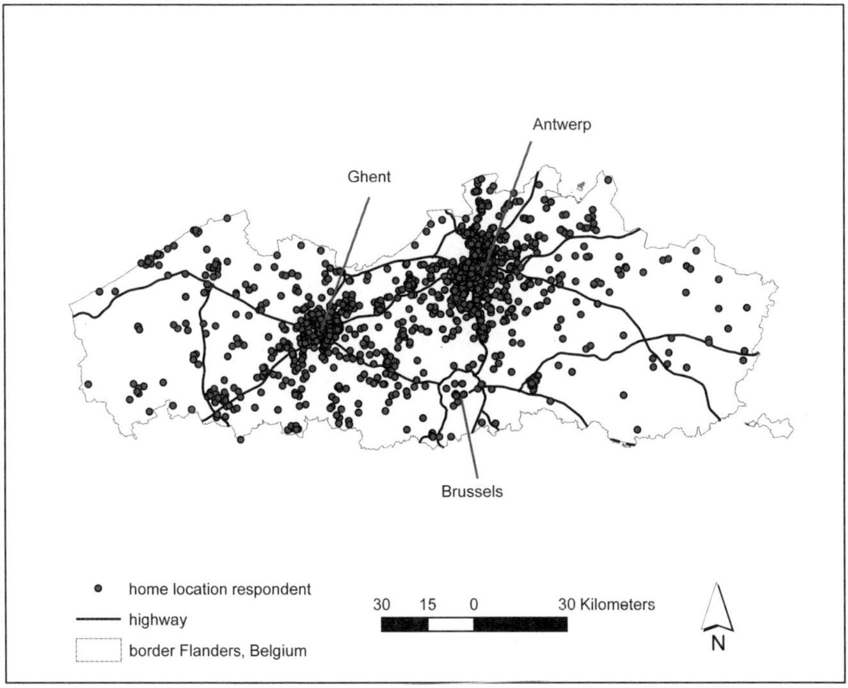


**Figure 3.1 Locations of respondents in Flanders**

## Spatial mismatch

This section describes how spatial mismatch is determined by balancing the spatial characteristics of the respondent's current residential neighbourhood against his/her subjective spatial perceptions.

### *Subjective spatial perceptions*

Although the survey was not designed to question perceptions, it contained 16 statements on how respondents perceive their current residential neighbourhood. Respondents were first asked to indicate the aspects that would influence a *supposed* residential location choice (such as quietness, presence of green areas, distance to work and traffic safety, but excluding price). Then, they were asked to indicate on a five-point Likert scale their level of satisfaction with these aspects rated important in their *current* residential neighbourhood. From these 16 statements, we selected only those seven statements that are relevant and can be related to the physical characteristics of the residential neighbourhood. They follow as: (i) closeness to public transport, (ii) closeness to shops and groceries, (iii) closeness to leisure activities, (iv) closeness to family and friends, (v) closeness to work, (vi) quietness and (vii) presence of green areas.

A correlation analysis revealed significant correlations between a large portion of statements on perceptions of the residential neighbourhood. This indicates that factor analysis might be useful to reduce the dimensionality and retrieve the underlying factors (Hair et al. 1998). The scores on these statements therefore went through factor analysis (principal axis factoring, promax rotation, 39.4 per cent variance explained) and were divided into two underlying dimensions influencing how respondents perceive their residence: (i) having access to various facilities (probably reflecting urban perceptions), and (ii) the presence of open space and quietness (probably reflecting suburban or rural perceptions). The number of factors in our analysis have been chosen based on interpretation of the scree plot, eigenvalues larger than one, and especially interpretation of the factors (for detailed information, see Van Acker et al. 2012).

In a subsequent step, respondents with similar scores on these two perception factors were grouped together by means of a cluster analysis (Ward's method, squared Euclidean distance). The number of clusters have been based on the interpretation of a graph in which the within-cluster sum of squares are plotted against the number of clusters (a sharp change may be indicative of the best solution) and, especially, the interpretability of the clusters. Doing so, we found three clusters reflecting whether respondents perceive their residential neighbourhoods as urban, suburban or rural. Urban perceptions are characterized by high ratings of accessibility and low ratings of open space and quietness, whereas the opposite holds for rural perceptions. A third cluster combines high ratings of accessibility with high ratings of open space and quietness. This refers to a combination of urban as well as rural perceptions, which we have labelled 'suburban'. This suburban perception obtains higher ratings of accessibility, open space and quietness compared to urban respectively rural perceptions, which might seem awkward. However, we should keep in mind that the ratings refer to *perceptions*, which do not mean that overall accessibility is better in suburban areas as compared to urban neighbourhoods, or that suburban areas have more open space than rural neighbourhoods. Furthermore, the survey questioned how *satisfied* respondents are with their current residential neighbourhood. It might be possible that suburban residents are more satisfied and enjoy the mix of positive urban and rural aspects (such as accessibility and, respectively, open space and quietness) to a greater extent than their urban and rural counterparts (for detailed information, see Van Acker et al. 2012).

### *Objective spatial characteristics*

By geocoding the respondent's address, we were able to add spatial information from various land use and transportation databases in order to calculate several spatial characteristics of the respondent's residence. For determining spatial mismatch, we calculated two additional spatial characteristics directly related to the seven statements on spatial perception of the residential neighbourhood: (i) local potential accessibility by car, and (ii) the built-up index.

Accessibility can be measured in various ways, but it always refers to the ability 'to reach activities or locations by means of a (combination of) travel mode(s)' (Geurs and van Wee 2004). We used a 'potential accessibility' measure, which is a simple but commonly-used accessibility measure. It calculates the number of activities that can be reached in a certain amount of time and weighs these calculations for travel time. We used the number of people that can be reached by car within 5 minutes as a proxy for local potential accessibility. For each residence, accessibility is calculated using the regional travel demand forecasting model Multimodal Model Flanders. It is basically the sum of the number of people of every census tract in the region, weighted by the travel time from the residence to these census tracts. Travel time is calculated in ArcGIS 9.2 as the fastest path by car along the road network. We restricted this travel time to 5 minutes in order to detect differences in local accessibility. After all, our study area has a limited geographical scale so that differences in accessibility are more important on a local level (for example, within 5 minutes) than a regional level (for example, within 60 minutes). Although we are aware that accessibility is more than just having access to people, we lacked detailed and geocoded information on, for example, the location of leisure activities, which would be more relevant for our analysis of mode choices for leisure trips. Consequently, we have limited our potential accessibility measure to having access to people. The built-up index equals the percentage of built-up surface at the census tract level. It can be considered as a proxy for built-up density and is derived from the land use database of the Agency of Spatial Information Flanders, which offers a categorization between built-up surfaces and open surfaces.

By performing a cluster analysis (Ward's method, squared Euclidean distance), neighbourhoods with similar scores on these two spatial characteristics are grouped together so that the clusters describe various residential neighbourhood types. The number of clusters are based on the interpretation of a graph in which the within-cluster sum of squares are plotted against the number of clusters (a sharp change may be indicative of the best solution) and especially, the interpretability of the clusters. Doing so, we found three clusters which also reflected a continuum ranging from urban, suburban and rural neighbourhoods. Urban neighbourhoods are characterized by high levels of accessibility (that is, many people can be reached by car within 5 minutes) and a high percentage of built-up area, whereas the opposite holds for rural neighbourhoods. A third cluster falls in-between, with moderate levels of accessibility and moderate percentages of built-up area. In other words, local accessibility and built-up density increase with increasing urbanization as can be expected (for detailed information, see Van Acker et al. 2012).

### *Spatial mismatch*

After having specified the respondents' spatial perceptions and the diverse neighbourhood types, we can balance these two against each other and determine whether respondents perceive their residence in a correct way.

Table 3.1 illustrates that almost 40 per cent of all respondents correctly perceive their residential neighbourhood (see the figures in grey, on the diagonal) and have what we call a spatial match. The large amount of spatial mismatch is thus striking. Moreover, respondents tend to overrate the urbanized character of their residence (see the larger figures below – the diagonal in grey compared to figures above this diagonal). For example, more than half of all respondents who reside in a rural neighbourhood perceive their residence as suburban (469 of the 814 rural respondents), whereas this figure is only 10 per cent in the reverse situation (that is, 53 of the 494 respondents residing in a suburban neighbourhood who perceive it as rural). This urbanized perception can be explained by the long-lasting tradition of suburbanization that exists in Belgium and goes back to the nineteenth century. After all, influenced by its housing policy and transport policy, a commuting culture has always existed in Belgium. Due to inexpensive public transport season tickets and a well-established network of railways and tramways, people were no longer compelled to reside close to jobs located within the city and they moved toward green, safe and quiet residential neighbourhoods outside the city centre. This was further encouraged by housing policy, which promoted inexpensive social house-construction in garden cities and provided subsidies and fiscal compensation for individual home-ownership. As a consequence, some form of suburbanization already existed in Belgium from the second half of the nineteenth century (Lauwers 1991, Kesteloot 2003, Verhetsel et al. 2007, Boussauw et al. 2009). This process of extensive suburbanization led to a highly fragmented urbanized space and has evoked the impression that every square metre is densely built-up.

**Table 3.1 Size of spatial (mis)match**

| **Perception cluster →**<br>**Spatial cluster ↓** | | **Urban** | **Sub-urban** | **Rural** | **Total** |
|---|---|---|---|---|---|
| **Urban** | N | 170 | 121 | 27 | **318** |
| | % of Total | 10.5% | 7.4% | 1.7% | **19.6%** |
| **Sub-urban** | N | 202 | 239 | 53 | **494** |
| | % of Total | 12.4% | 14.7% | 3.3% | **30.4%** |
| **Rural** | N | 138 | 469 | 207 | **814** |
| | % of Total | 8.5% | 28.8% | 12.7% | **50.1%** |
| **Total** | **N** | **510** | **829** | **287** | **1,626** |
| | **% of Total** | **31.4%** | **51.0%** | **17.7%** | **100.0%** |

One might argue that any association between spatial mismatch and mode choice is the result of various sample biases (for example, the overrepresentation of highly educated people). However, we found no statistically significant socio-economic and demographic differences to exist between matched and mismatched respondents (for detailed information, see Van Acker et al. 2012).

## Spatial mismatch and its travel consequences

### *Descriptive analysis*

Respondents were also asked what kind of leisure trips they perform on a monthly basis and which travel mode they generally use for this. In a previous study (Van Acker et al. 2012) we described how these mode choices for leisure activities differ between respondents with (mis)matched spatial perceptions. We hypothesized that if spatial perceptions are crucial to mode choices, then it would be plausible that respondents with mismatched perceptions choose travel modes corresponding to their spatial perceptions. However, if perceptions are not crucial to mode choices, the influence of the residential neighbourhood itself might become more important, so that all inhabitants within a particular neighbourhood type would make similar mode choices regardless of any (mis)matched spatial perception. In our analysis, we found evidence for both hypotheses.

For example, residing in an urban neighbourhood seems to discourage car use (left side of Figure 3.2). Car use is almost equally high for all respondents residing in an urban neighbourhood. Whether someone perceives this neighbourhood as urban or not, it does not seem to influence the decision to use the car. However, this does not hold for a suburban or rural neighbourhood. Here, perceptions become more important. A suburban resident who perceives their residence as urban (or rural), tends to act as a matched urbanite (or matched 'ruralite'), therefore using the car less often (or more often).

At first sight it also seems that an urban neighbourhood encourages the use of public transport (centre of Figure 3.2) and walking/cycling (right side of Figure 3.2). Even though some mismatched urbanites perceive their urban residence as suburban, they tend to behave as matched urbanites and are more likely to use public transport, and walk/cycle more often than they actually would do so by virtue of their spatial perception. This association is less clear for mismatched urbanites who perceive their neighbourhood as rural (instead of urban). Their share of public transport and walking/cycling is lower than that of a matched urbanite (suggesting that it is not only about the spatial environment), yet still considerably higher than a matched ruralite (suggesting that perceptions are not the only influences either). More or less similar patterns are found for rural dwellers, but mode choices of suburban dwellers are possibly influenced more by spatial perceptions than by the suburban neighbourhood itself. Mismatched suburbanites tend to make similar mode choices to their matched counterparts. For example, evidence suggests that someone who

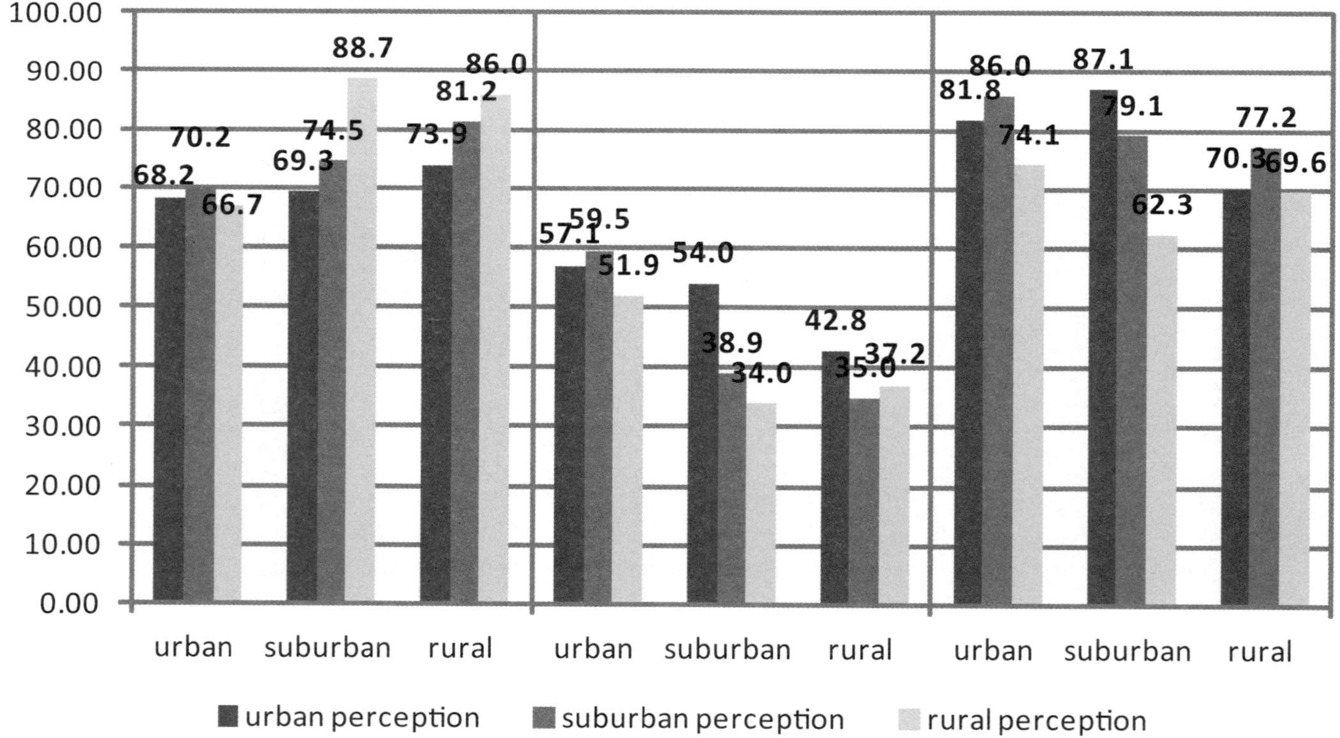


**Figure 3.2** **Influence of (mis)matched spatial perceptions on mode choices: car use (left side), public transport (centre), walking/cycling (right side)**

perceives their suburban residence as urban (or rural) also behaves as a matched urbanite (or ruralite) and chooses more frequently (or less frequently) to use public transport and to walk/cycle.

*Hierarchical logistic regression analysis*

The descriptive analysis in the section 'Spatial mismatch and its travel consequences' depicts a possible *association* between mode choice for leisure activities and (mis)matched spatial perceptions. However, this interesting yet initial association should be controlled for socio-economic and demographic influences. After all, association is not the same as *causation*. For example, previous research has found that car ownership mediates the relationship between the built environment, attitudes and mode choices (Van Acker and Witlox 2010, Van Acker et al. 2011). Our first impression based on Figure 3.2 might thus be biased because of a possible interaction with car ownership. Therefore, we have extended our previous study by statistically testing the possible effect of spatial (mis)match on mode choices for leisure activities in order to control for differences in socio-economic and demographic background variables. This has been done by means of a hierarchical regression analysis. The difference with a regular regression analysis is that in a hierarchical regression analysis, independent variables are entered in blocks. Independent variables that need to be controlled are usually first included in the regression model.

In this analysis, we are especially interested in the effect of the built environment and spatial (mis)match on mode choices. At the same time though, we must also be aware that other variables like car ownership might be associated with the built environment, spatial (mis)match and mode choice. To make sure that car ownership does not strongly impact the entire association between the built environment and spatial (mis)match or mode choice, we have put variables such as car ownership first in the regression model. This has ensured that controlling variables such as car ownership will get 'credit' for any shared variability they may have with the predictors we are really interested in, especially the built environment and spatial (mis)match. Any observed effects of these predictors can then be said to be 'independent of' the effects of other variables already controlled for. This procedure thus allows us to test the effects of certain predictors independent of the influence of others. Since car use, public transport and cycling/walking are defined every time in our study as a binary variable, we have chosen to estimate three separate hierarchical logistic regression analyses in SPSS Statistics 20.

In this analysis, we controlled the interaction between the built environment, spatial (mis)match and mode choice for the following personal characteristics: gender, marital status, the presence of young children below 12 years of age in the household, education, employment status, household income, possession of a driving license and car ownership. These personal characteristics were entered in the first block of the three hierarchical logistic regression analyses. Moreover, these controlling variables were entered into the model by using the conditional forward method. This means that the procedure started with no variables in the model, tested the addition of each variable using a chosen model comparison criterion, added the variable (if any) that improves the model the most, and repeated this process until no further improvements could be made to the model.

Residential neighbourhood types (with three categories – urban, suburban, rural as described in the section 'Objective spatial characteristics') were entered in a second block. Spatial (mis)match was entered in a final third block. The results of Table 3.1 were translated into a new categorical variable measuring spatial (mis)match by three categories: (i) respondents with correct spatial perceptions (corresponding to the figures on the diagonal in grey in Table 3.1), (ii) respondents who overrate the urbanized character of their residence (corresponding to the figures below the diagonal in Table 3.1) and (iii) respondents who underrate the urbanized character of their residence (corresponding to the figures above the diagonal in Table 3.1).

The results of the three hierarchical logistic regression analyses are summarized in Table 3.2. We are primarily interested here in the effect of the built environment on mode choices for leisure activities compared to the effect of spatial (mis)match. Therefore, the discussion of the results will mainly focus on these effects.

Given the same personal background variables, respondents residing in a suburban and rural neighbourhood are more likely to use their cars for leisure activities compared to respondents residing in an urban neighbourhood. The opposite holds for public transport and cycling/walking. Spatial (mis)match also has the expected effect on mode choice. When compared to respondents with correct spatial perceptions of their residence, respondents overrating the urbanized character of their residence tend to use their cars less frequently and to use public transport and cycle/walk more frequently. The opposite holds for respondents underrating the urbanized character of their residence (more car use, less public transport and less cycling/walking). This seems to confirm our finding in the descriptive analysis that the built environment as well as spatial perceptions both influence mode choices for leisure activities.

However, the effects of residential neighbourhood type and spatial (mis) match are in many cases not statistically significant. Whereas the descriptive analysis suggests that car use is influenced by the built environment (especially for residents of an urban neighbourhood) as well as spatial perceptions (especially for residents in a suburban or rural neighbourhood), the hierarchical logistic regression analysis points out that no significant effects exist from both. On the contrary, car ownership seems to be the most important predictor of car use for leisure activities (as well as for other mode choices). The Wald statistic has the largest value for car ownership indicating that car ownership contributes the most to the regression model. Contrary to car use, the residential neighbourhood type has a significant effect on the use of public transport and cycling/walking. The effect of residing in a rural neighbourhood compared to an urban neighbourhood was especially found to be significant. The question remains to which degree public transport and cycling/walking are influenced by the way people perceive their residence. Spatial (mis)match was found to only be statistically significant for cycling/walking, especially for residents underrating the urbanized character of their residence compared to residents who perceive their residence correctly.

Table 3.3 summarizes different model fit measures for the three hierarchical logistic regression analyses. The results of the Chi²-test illustrate that all three analyses (car use, public transport, cycling/walking) obtain a good model fit: no significant differences exist between the observed and estimated data. This Chi²-test compares the log likelihood of the estimated model (831.533 for car use, 1199.614 for public transport and 970.314 for cycling/walking) with the log likelihood of the model without any predictors. Logistic regression analysis does not report the proportion of explained variance ($R^2$) such as for an OLS linear regression analysis. However, some pseudo $R^2$-measures are reported. Cox and Snell $R^2$ and Nagelkerke $R^2$ increases, although slightly, for each block of predictors that is added to the hierarchical logistic regression models. Residential neighbourhood type and spatial (mis)match thus adds some explanatory power to the models, but its influences remain limited.

**Table 3.2 Three hierarchical logistic regression analyses of mode choices for leisure activities**

| | B | Std. Error | Wald | Sig. | Exp (B) |
|---|---|---|---|---|---|
| **CAR USE** | | | | | |
| constant | –0.733 | 0.294 | 6.223 | 0.013 | 0.481 |
| female | 0.829 | 0.182 | 20.728 | 0.000 | 2.290 |
| no driving license | –1.115 | 0.204 | 29.840 | 0.000 | 0.328 |
| unemployed (ref.) | | | 16.631 | 0.000 | |
| part-time employed | 0.337 | 0.330 | 1.045 | 0.307 | 1.401 |
| full-time employed | 0.864 | 0.212 | 16.625 | 0.000 | 2.372 |
| car ownership | 1.138 | 0.131 | 75.273 | 0.000 | 3.121 |
| urban neighbourhood (ref.) | | | 0.532 | 0.767 | |
| suburban neighbourhood | 0.106 | 0.266 | 0.158 | 0.691 | 1.112 |
| rural neighbourhood | 0.199 | 0.276 | 0.518 | 0.472 | 1.220 |
| spatial match (ref.) | | | 2.519 | 0.284 | |
| overrated | –0.193 | 0.205 | 0.891 | 0.345 | 0.824 |
| underrated | 0.274 | 0.272 | 1.014 | 0.314 | 1.315 |
| **PUBLIC TRANSPORT** | | | | | |
| constant | 1.213 | 0.330 | 13.499 | 0.000 | 3.364 |
| age | –0.016 | 0.009 | 3.164 | 0.075 | 0.985 |
| female | 0.685 | 0.147 | 21.803 | 0.000 | 1.985 |
| no driving license | 0.408 | 0.189 | 4.645 | 0.031 | 1.503 |
| children -12 years | –0.523 | 0.205 | 6.505 | 0.011 | 0.593 |
| unemployed (ref.) | | | 15.783 | 0.000 | |
| part-time employed | –0.890 | 0.314 | 8.015 | 0.005 | 0.411 |

| | B | Std. Error | Wald | Sig. | Exp (B) |
|---|---|---|---|---|---|
| full-time employed | –0.731 | 0.191 | 14.626 | 0.000 | 0.481 |
| car ownership | –0.459 | 0.093 | 24.315 | 0.000 | 0.632 |
| urban neighbourhood (ref.) | | | 7.483 | 0.024 | |
| suburban neighbourhood | –0.312 | 0.222 | 1.977 | 0.160 | 0.732 |
| rural neighbourhood | –0.615 | 0.231 | 7.110 | 0.008 | 0.541 |
| spatial match (ref.) | | | 0.445 | 0.800 | |
| overrated | 0.098 | 0.161 | 0.370 | 0.543 | 1.103 |
| underrated | –0.026 | 0.222 | 0.013 | 0.908 | 0.975 |
| **CYCLING / WALKING** | | | | | |
| constant | 3.560 | 0.369 | 93.076 | 0.000 | 35.180 |
| age | –0.017 | 0.009 | 3.791 | 0.052 | 0.983 |
| unemployed (ref.) | | | 15.354 | 0.000 | |
| part-time employed | –0.208 | 0.323 | 0.415 | 0.520 | 0.812 |
| full-time employed | –0.756 | 0.209 | 13.152 | 0.000 | 0.469 |
| car ownership | –0.483 | 0.103 | 21.742 | 0.000 | 0.617 |
| urban neighbourhood (ref.) | | | 8.811 | 0.012 | |
| suburban neighbourhood | –0.317 | 0.275 | 1.326 | 0.250 | 0.728 |
| rural neighbourhood | –0.811 | 0.295 | 7.569 | 0.006 | 0.444 |
| spatial match (ref.) | | | 10.213 | 0.006 | |
| overrated | 0.014 | 0.183 | 0.006 | 0.940 | 1.014 |
| underrated | –0.811 | 0.265 | 9.358 | 0.002 | 0.444 |

**Table 3.3 Model summary statistics**

| | Chi$^2$ (df) p | -2 Log Likelihood | Cox & Snell R$^{22}$ | Nagelkerke R$^2$ |
|---|---|---|---|---|
| **CAR USE** | | | | |
| block 1 | 209.089 (5) 0.000 | 834.135 | 0.190 | 0.292 |
| block 2 | 209.119 (7) 0.000 | 834.105 | 0.190 | 0.292 |
| block 3 | 211.670 (9) 0.000 | 831.553 | 0.192 | 0.295 |
| **PUBLIC TRANSPORT** | | | | |
| block 1 | 157.277 (7) 0.000 | 1208.501 | 0.146 | 0.196 |
| block 2 | 165.719 (9) 0.000 | 1200.059 | 0.154 | 0.206 |
| block 3 | 166.164 (11) 0.000 | 1199.614 | 0.154 | 0.206 |
| **CYCLING / WALKING** | | | | |
| block 1 | 62.712 (4) 0.000 | 983.057 | 0.061 | 0.094 |
| block 2 | 65.219 (6) 0.000 | 980.550 | 0.064 | 0.098 |
| block 3 | 75.455 (8) 0.000 | 970.314 | 0.073 | 0.112 |

## Conclusions

This paper contributes to research on the interaction between the built environment and travel behaviour by evaluating the objective and subjective spatial influences on mode choices for leisure activities. Recent land use-travel behaviour interaction studies are aware of the influence of subjective aspects such as perceptions in addition to the objectively measured characteristics of the built environment. Nevertheless, these studies tend to neglect the question of whether these subjective influences correspond to the objective reality. Therefore, this paper aims at describing the size of spatial (mis)match between perceptions and reality.

Spatial mismatch occurs to a large degree in Flanders. Only 40 per cent of all respondents perceive their residence in a correct way. Moreover, due to the long-lasting tradition of suburbanization, which has resulted in the ubiquitous impression of Flanders as one densely built-up area, many respondents tend to overrate the urbanized character of their residential neighbourhood. Based on a descriptive analysis, one might conclude that these spatial perceptions are crucial to mode choice for leisure activities, especially for suburbanites. Among all suburbanites, public transport, cycling and walking (or car use) are highest among mismatched suburbanites perceiving their residence as urban (or rural). Within the suburbs, it seems that residents are able to choose travel modes that fit within their perception of the residence. However, the results of three hierarchical logistic regressions reveal that hardly any significant effect of spatial (mis)match can be found on mode choices for leisure activities. Only cycling and walking were statistically influenced by the spatial perceptions.

Spatial perceptions are not the only predictors of mode choices. The descriptive analysis suggested that in some cases the residential neighbourhood itself becomes more important. Especially in urban neighbourhoods, it seems that high densities and high local accessibility almost automatically result in a lower car share, a higher public transport share and more walking and cycling. Differences in how respondents perceive their urban residence seemed less important though, where matched and mismatched urbanites tend to make similar mode choices. As mentioned earlier though, the influence of residential neighbourhood type was found significant only for the use of public transport as well as cycling and walking.

Consequently, the results of the hierarchical logistic regressions illustrate that the association between mode choice and (mis)matched spatial perceptions depicted in the descriptive analysis does not automatically translate into causation. When controlling for socio-economic and demographic characteristics, it seems that the built environment and perceptions add only a small amount of explanatory power to the models. The question thus remains of which mechanisms underlie the travel decision process. Is it really a question of balancing the spatial characteristics with personal perceptions while controlling for socio-economic and demographic background variables? Or, to what extent are mode choices for leisure activities also irrational decisions determined by, for example, habits repeated over time, impulsive decisions or a general lifestyle? More research is thus needed to disentangle causation from association so that a behavioural approach can be elaborated in land use-travel behaviour interaction research.

Based on our findings, important recommendations can be made for spatial planning policies. Spatial planning policies aimed at densifying and providing facilities at the neighbourhood level can contribute to a more sustainable mobility. The results of the three hierarchical logistic regression analyses indicate that such spatial planning policies would significantly result in more use of public transport and more cycling/walking, but not in less car use. Car use is mainly determined by car ownership. Besides making cars directly more expensive to own and operate (such as through registration fees, gasoline taxes and road pricing), spatial planning policies should develop residential neighbourhood types that do not require the ownership of a car. This should further be supplemented by general promotional campaigns to raise awareness among people that owning and driving a car is not something as self-evident as it seems today. Campaigns should promote the attractiveness of other modes such as public transport, cycling and walking, while also promoting the advantages of living in high-density, diverse and accessible urban neighbourhoods. In sum, an urban revival is still needed. This need is also supported by our finding that not everyone will use public transport or will cycle or walk more frequently when faced with increased densities and levels of local accessibility. Such planning policies will remain unsuccessful for respondents who underrate the urbanized character of their residence and who thus misperceive their residence as non-urban; underrating the urbanized character of the residence results in less cycling and walking.

The findings of this chapter illustrate that the effect of spatial planning policies can only be correctly understood when subjective influences such as perceptions are also accounted for. Otherwise, one might have expectations that are too high. Furthermore, the fact that car use for leisure activities remains high, even in urban neighbourhoods, suggests that car ownership has an important role in mode choices. Once people own a car, they tend to use it more often. Urban planning policies should therefore not only focus on influencing mode choices directly by measures of increasing density and diversity, but also indirectly by measures related to car ownership.

## References

Bagley, M.N. and Mokhtarian, P.L. 2002. The impact of residential neighborhood type on travel behavior: a structural equations modeling approach. *Annals of Regional Science*, 36 (2), 279–97.

Bartholomew, K. and Ewing, R. 2009. Land use-transportation scenarios and future vehicle travel and land consumption. *Journal of the American Planning Association*, 75 (1), 13–27.

Boussauw, K., Zwerts, E. and Witlox, F. 2009. Mobiel Vlaanderen, in *De sociale staat van Vlaanderen 2009*, edited by L. Vanderleyden et al. Brussel: Studiedienst van de Vlaamse Regering, 279–312 [in Dutch].

Geurs, K. and van Wee, B. 2004. Accessibility evaluation of land-use and transport strategies: review and research directions. *Journal of Transport Geography*, 12 (2), 127–40.

Hair, J.F. Jr., Anderson, R.E., Tatham, R.L. and Black, W.C. 1998. *Multivariate Data Analysis*. New Jersey: Prentice Hall.

Handy, S. 2002. Travel behaviour – land use interactions: an overview and assessment of the research, in *In Perpetual Motion: Travel Behavior Research Opportunities and Application Challenges*, edited by H.S. Mahmassani. Amsterdam: Pergamon, 223–36.

Handy, S. 2005. *Critical Assessment of the Literature on the Relationships among Transportation, Land Use and Physical Activity*. Washington D.C.: Transportation Research Board.

Kesteloot, C. 2003. Verstedelijking in Vlaanderen: Problemen, kansen en uitdagingen voor het beleid in de 21ste eeuw, in *De eeuw van de stad*, edited by L. Boudry et al. Brussel: Ministerie van de Vlaamse Gemeenschap, 15–39 [in Dutch].

Kitamura, R., Mokhtarian, P.L. and Laidet, L. 1997. A micro-analysis of land use and travel in five neighborhoods in the San Francisco Bay Area. *Transportation*, 24 (2), 125–58.

Lauwers, D. 1991. Ruimtelijke aspecten van de fileproblematiek, in *Files in Vlaanderen*, edited by R. Poté. Leuven: Garant, 48–60 [in Dutch].

Scheiner, J. and Holz-Rau, C. 2007. Travel mode choice: affected by objective or subjective determinants? *Transportation*, 34 (4), 487–511.

Schwanen, T. and Mokhtarian, P.L. 2005a. What if you live in the wrong neighborhood? The impact of residential neighborhood type dissonance on distance traveled. *Transportation Research D*, 10 (2), 127–51.

Schwanen, T. and Mokhtarian, P.L. 2005b. What affects commute mode choice: neighborhood physical structure or preferences toward neighborhoods? *Journal of Transport Geography*, 13 (1), 83–99.

Van Acker, V., Derudder, B. and Witlox, F. 2012. Why people use their cars while the built environment imposes cycling. *Journal of Transport and Land Use* (forthcoming).

Van Acker, V., Mokhtarian, P.L. and Witlox, F. 2011. Going soft: on how subjective variables explain modal choices for leisure travel. *European Journal of Transport and Infrastructure Research*, 11 (2), 115–46.

Van Acker, V., van Wee, B. and Witlox, F. 2010. When transport geography meets social psychology: toward a conceptual model of travel behaviour. *Transport Reviews*, 30 (2), 219–40.

Van Acker, V. and Witlox, F. 2005. Exploring the relationships between land use system and travel behaviour concepts: some first findings. *Solstice*, 16 (1), 1–18.

Van Acker, V. and Witlox, F. 2010. Car ownership as a mediating variable in car travel behaviour research using a structural equation modelling approach to identify its dual relationship. *Journal of Transport Geography*, 18 (1), 65–74.
van Wee, B. 2002. Land use and transport: research and policy challenges. *Journal of Transport Geography*, 10 (4), 259–71.
van Wee, B., Holwerda, H. and van Baren, R. 2002. Preferences for modes, residential location and travel behaviour: the relevance for land-use impacts on mobility. *European Journal on Transport and Infrastructure Research*, 2 (3/4), 305–16.
Verhetsel, A., Thomas, I. and Beelen, M. 2007. De kracht van het Alonso-Mutch model. Het woon-werkverkeer in en rond de Belgische grootsteden. *Ruimte en Planning*, 27 (3), 9–28 [in Dutch].

Chapter 4

# Gender Roles between Traditionalism and Change: Time Use for Out-of-Home Activities and Trips in Germany, 1994–2008

Joachim Scheiner

## Introduction

For about three decades, gender relations have been an important aspect of time use and transport studies. In recent years, a research field has emerged within this area that focuses on interdependencies in activity and trip patterns between partners in a household (see Ettema and Van der Lippe 2009). Although this issue is clearly related to gender, in many cases research tends to focus on improving activity-based simulations of travel behaviour rather than on developing a better understanding of gender relations.

In this chapter, the distribution of time used for activities and trips by women and men living as a couple or in a family household is studied using data collected in Germany. Cross-sectional analyses as well as trends during the period of 1994 to 2008 are presented. This research focuses on using descriptive statistics to show how many households exercise various types of work-sharing (seen as traditional, egalitarian and reversed role) and investigates the factors impacting the division of work in these households. To my knowledge, this is the first time the distribution of time use in partnerships (rather than individual mean values) has been studied from a transport research perspective.

The chapter hence focuses on mobility in terms of travel time plus out-of-home activity time studied from a gender perspective. In the field of sustainability, gender equity in time use, travel and activity patterns has been identified as important, not only because it may help understand gendered effects of transport on society, the economy and the environment, but also because equity itself is a key element of sustainability (Root et al. 2000, Hanson 2010).

This chapter is informed by several disciplines. From a theoretical perspective (see section on theoretical background) this study draws on gender studies related to time budgets and travel. Time geography provides an analysis framework for the temporal structure of daily life in an individual's social context. Empirically, this study is informed by research on time use and travel behaviour (see section on empirical studies). The data and methodology are introduced in the section on methodology. The findings are presented in the section on results. In the last section, some conclusions are drawn and an outlook on further research is given.

## State of the research

### *Theoretical background*

Theoretical approaches to explaining gender differences in both travel behaviour and time use, including the gendered division of work, may be traced back to four key hypotheses, which are briefly summarized as follows (see for discussion Hanson and Pratt 1988, Law 1999, Uteng and Cresswell 2008, Scheiner and Holz-Rau 2012). The first looks at women's limited access to economic, social and temporal resources. The second, referred to as the 'household responsibility hypothesis', focuses on the social roles that may commit women to take on household and family responsibilities more than men. The third looks at how gender attitudes and norms based on preferences may cause gender differences. Finally, the fourth hypothesis focuses on the patriarchal power relations that may operate over and above all other inequalities and differences.

All of these approaches share a common understanding of paid, marketed work (employment), which is primarily undertaken by men and serves to promote male economic and social dominance. This perspective is typically seen in opposition to unpaid, non-market, household work, which is primarily undertaken by women and therefore limits their access to the public sphere and economic independence.

Methodologically, travel and activity studies are deeply rooted in time geography (Hägerstrand 1970). Time geography analyses individual or household patterns of time use, trip and activity sequences in the context of three types of constraints: capability, authority and coupling. Capability constraints mainly reflect a person's individual requirements and limitations (eating, sleeping, limited resources), while authority and coupling constraints are imposed on an individual by others.

Coupling constraints are of particular interest here, because household maintenance and income generation are normally interdependent between partners. When somebody goes shopping, they would normally serve their partner as well. Further, when a member of the household is employed, the earnings would normally serve to improve the household budget. In this household-based logic, time use for unpaid household duties and paid employment depends on the partner's time use. Such coupling constraints reflect intra-household negotiations about time and other resources, such as cars (Scheiner and Holz-Rau 2012). The result of such negotiations may be interpreted in terms of power relations, especially who prevails, or household economy, otherwise known as attempts to efficiently balance resource input and benefit for the household as a whole.

Gender relations have changed considerably in recent decades. Beck (1992) has elaborated extensively on the dissolution of gender role expectations and the making of space for permanently negotiable gender roles. The 'male breadwinner and female housewife' model has not vanished, but it has become more of a rarity than an everyday norm. New household types have also emerged and 'old' household types have changed in quantity (Wagner and Franzmann 2000). At the same time, the level of professional qualification in women has increased, and new forms of employment have emerged or become more prevalent. Such changes affect activity patterns as well as travel behaviour.

### *Empirical studies*

The trends outlined above have developed hand in hand with an increase in both female employment and male housework. Related studies are so numerous that this section is therefore limited to key findings. Our focus is on Germany, and for this the most important publication is an edited volume (see Statistisches Bundesamt 2004) focusing on changes during the period of 1991/92 to 2001/02. Cross-national comparisons can also be found in BMFSFJ/Statistisches Bundesamt (2003).

Gille and Marbach (2004) studied trends in time use of couples during the time period above. They found that time expenditure for work *in total* (paid plus unpaid) hardly differs between genders. In all studied sub-groups (1991 and 2001, couples with and without children), work time expenditure was somewhat higher for men than for women. More distinct differences emerged when a difference was made between paid and unpaid work. Time use for unpaid work among men increased during the study period, as did time use for paid work among women. The relative overload women have in terms of unpaid work (odds ratios women/men) consequently declined considerably from factor 1.7 in 1991 to 1.4 in 2001. Virtually no changes were detected among parents though. Mothers in the study made a constantly above-average contribution to unpaid work. There is more equity between the genders in East Germany compared to West Germany (see also Cooke 2006).

Such trends have been observed to a similar extent across various industrialized countries (Bianchi et al. 2000, Blau et al. 2010). Nevertheless, gender specifics in the division of work still exist and are also similar across various countries (Baxter 1997). On average, women have a more tumultuous daily life than men, particularly when raising infants. This is often reflected in the mothers' more complex trip chains (McGuckin and Murakami 1999).

Other related studies tend to focus on comparing gender using mean values and proportions. Interactions in time use and activity patterns between partners in a household are studied mostly on the basis of complex modelling (Ettema and Van der Lippe 2009). While these studies provide much insight into relationships within households, household typologies according to the share of work within a household as well as gender role combinations are rare. Those that are found tend to exist within the fields of family sociology and psychology.[1]

1 This lack of research in time budget and transport studies is probably due to the structure of data used, which is typically limited to single random days that do not permit the construction of typologies.

Grunow et al. (2007) studied the allocation of housework between married couples within German households. They found that at the time of marriage, more than 40 per cent of couples had a cooperative, egalitarian work-sharing arrangement and less than 2 per cent of all couples had reversed roles. However, cooperative arrangements decline steeply afterwards.

Marks et al. (2009) studied gender role attitudes in US families, focusing on both parents and children. They identified three combinations of attitudes: (1) parents and children who are egalitarian; (2) parents and children who are traditional; and (3) parents who are traditional, but with egalitarian children. Most family conflicts were found to appear in the second group.

Overall, household work-sharing arrangements can be categorized into traditional, egalitarian and reversed role models. There are clear differences between genders in terms of participation in employment and household chores. Such differences not only apply to men and women in general, but also to partners in a household, particularly in families. Over time differences seem to decline in various spatial contexts.

## Methodology

### *Data*

Data used in this chapter are from the German Mobility Panel (GMP) and stem from the period of 1994 to 2008. Every household in this study was surveyed three times over a period of three years (Chlond and Kuhnimhof 2005).[2] A trip diary was used to collect information on trips throughout a complete week from all household members aged ten years or older. Additionally, socio-demographic attributes for the household and its members were collected as well as spatial context attributes of the residence and of the household members' workplaces or places of education.

Three important data limitations have to be kept in mind. First, the data do not include information on gender attitudes that may affect household work-sharing arrangements. Second, as in most travel data, in-home activities are not included. Accordingly, the scope of this chapter is limited to out-of-home activities. This is particularly important for household work that is to a large extent done at home. Third, data was limited to one week of reporting. Even if a trip diary over one week might be considered a good data base, untypical weeks can affect the results, such as when a wife becomes ill and her husband spends this time shopping and escorting the children.[3]

2 The GMP is a household survey conducted by the University of Karlsruhe on behalf of the Federal Ministry of Transport, Building and Urban Development (BMVBS). Data have been provided for research use by the Clearingstelle Verkehr (www.clearingstelle-verkehr.de).

3 The data include a self-reported measure of whether or not the week of report was 'more or less as always'. When the analysis is limited to these 'normal' weeks, traditional

*Analysis*

Analysis required significant decisions and data transformations in four respects: (1) the determination of trip purposes, (2) the measurement of target variables, (3) the measurement of explanatory variables for modelling and (4) the determination of household types and decisions about partners.

First, trip purposes have been determined using a mixed trip/tour-based concept. In simple trip chains (outward plus return trip), both trips are assigned to the activity at the destination. In complex chains with job-related or education trips that include an additional activity, the longest trip in the part of the chain is assigned to the job purpose (or educational purpose, as appropriate) while the shorter trip is assigned to the additional activity. For example, when a respondent goes shopping close to his/her workplace after work, and thus afterwards makes the (relatively long) trip home, the trip home from shopping would be classified as a job trip as this is the 'actual' commute. The short trip from the workplace to the grocery store would be treated as the shopping trip.

The use of this scheme results in a few cases of multiple purposes (1.3 per cent of all trips). In these cases the decision about the main purpose is made by using the following hierarchy: business, job, education, escort, shopping/errand, leisure or other.

The second type of data transformation includes participation in household and employed work, which is measured by time expenditure. Here, days without trip records are excluded, because reasons for this cannot be conclusively determined. Job and business activities/trips are categorized as employment, while shopping/errands and escort activities/trips are categorized as household work. Activities following an escort trip are categorized as household work. Therefore, it is assumed that an escort trip is typically followed by household (or family) work, such as waiting to pick up an elderly family member or attending a child's birthday party, even if an escort trip may be followed by a leisure activity. However, as Table 4.1 shows, escort activity time has only a minor effect on household work time.

The extent of traditionalism in a household is studied with respect to the division of labour between the partners. Two variables are used:

1. The proportion of the woman's time expended on household work (activities plus trips) in two partners' total time expenditure for household work ('traditionalism in household work');

work-sharing is indeed somewhat more frequent, particularly among households with children aged 13 or less (table available from the author upon request). The difference to the findings reported here, however, is very limited. Anyway, researchers should not focus analysis on 'normal' weeks or days. 'Non-normality', irregularities and non-routine all belong to normal daily life, particularly because normality only accounts for a relatively small share of real-world experience. Both partners identify their report week as having been 'more or less as always' in not more than 51 per cent of the households studied here.

2. The proportion of the man's time expended on employment (activities plus trips) in both partners' total time expenditure for employment ('traditionalism in employment').

For the construction of household types, a threshold value that determines which degree of similarity is considered (approximately) equal has to be defined. There is no theoretical and little empirical foundation for this (for example, Marks et al. 2009, Grunow et al. 2007). Using the income thresholds defined by Grunow et al. (2007: 170), two partners' time expenditure is considered equal as long as both of the two contribute between 40 per cent and 60 per cent to total household work or employed work respectively. If the male partner contributes less than 40 per cent to household work, or the female partner contributes less than 40 per cent to employment, this arrangement is treated as 'traditional'. Shares of less than 20 per cent are treated as 'strongly traditional'.

(3) Socio-demographics, as well as spatial, travel and time budget attributes, are used as explanatory variables in regression models. Tables 4.1 and 4.2 show descriptive statistics and detailed definitions of the variables used. The regressions are limited to households with at least one employed partner.

**Table 4.1 Frequency of categorical variables (used as dummy variables in regression models)**

| | **Households** |
|---|---|
| Age category | |
| Both partners 18–39 Jahre | 28.5 |
| M 18–39, F 40–59 Jahre | 2.2 |
| M 40–59, F 18–39 Jahre | 11.5 |
| M 40–59, F 40–59 Jahre | 44.8 |
| M < 60, F 60+ Jahre | 1.2 |
| M 60+, F < 60 Jahre | 6.6 |
| M 60+, F 60+ Jahre | 5.2 |
| n | 4,507 |
| Relative education level* | |
| Both partners have same education level | 52.9 |
| M: dominates strongly | 5.6 |
| M: dominates somewhat | 21.3 |
| F: dominates somewhat | 16.5 |
| F: dominates strongly | 3.7 |
| n | 4,413 |
| Education level of household | |
| Both partners elementary school or no degree | 7.0 |
| At least one partner has elementary school degree with apprenticeship | 17.2 |
| At least one partner has secondary school level I | 30.3 |

| | |
|---|---|
| At least one partner has final secondary school degree (university entrance) or university degree | 45.5 |
| n | 4,700 |
| Relative employment level** | |
| Both partners employed to the same extent | 26.1 |
| M: dominates strongly | 25.2 |
| M: dominates somewhat | 37.1 |
| F: dominates somewhat | 7.1 |
| F: dominates strongly | 4.5 |
| n | 4,503 |
| Employment level of household: at least one partner is full-time employed | |
| Yes | 86.1 |
| No | 13.9 |
| n | 4,711 |
| Municipality size category | |
| < 20,000 inh | 40.2 |
| 20,000–100,000 inh | 26.6 |
| 100,000–500,000 inh | 18.4 |
| > 500,000 inh | 14.8 |
| n | 8,240 |
| Central residential location within municipality*** | |
| Yes | 19.6 |
| No (or municipality < 20,000 inh) | 80.4 |
| n | 8,363 |
| East Germany | |
| Yes | 17.8 |
| No | 82.2 |
| n | 8,364 |

F: female, M: male.

* Education level is measured as a five-point ordinal scale: no degree, elementary school without apprenticeship, elementary school with apprenticeship, secondary school degree, university entrance or university degree. 'Dominating somewhat' means that the respondent's achievement is one point higher on this scale than his/her partner's degree, and (s)he has achieved at least elementary level with apprenticeship. 'Dominating strongly' means that the respondent's achievement is at least two points higher than his/her partner's degree.

** Employment is measured in three categories: full-time, part-time, not employed. Trainees are treated as full-time employed, as similar time constraints apply. 'Dominating strongly' means that the respondent is full-time employed, while his/her partner is unemployed. Other cases of inequality (couples working full-time plus part-time or part-time plus unemployed) are coded as 'dominating somewhat'.

*** The location within the municipality (central vs. remote) is based on respondents' self-report. This attribute is available in municipalities with more than 20,000 inhabitants. Smaller municipalities are coded as 'remote', because the central areas in large cities are known to be particularly specific in social terms.

**Table 4.2 Descriptives of cardinal-level variables used in regression models**

| | Minimum | Maximum | Mean | Standard deviation | Skewness | Kurtosis | n |
|---|---|---|---|---|---|---|---|
| Number of children < 10 years in household | 0 | 5 | 0.34 | 0.69 | 2.04 | 3.60 | 8,340 |
| Number of children 10–13 years in household | 0 | 3 | 0.13 | 0.38 | 3.18 | 10.41 | 8,364 |
| Number of children 14–17 years in household | 0 | 3 | 0.13 | 0.38 | 3.01 | 9.19 | 8,364 |
| Car availability (number of cars minus number of licensed household members) | –4 | 3 | –0.27 | 0.71 | 0.11 | 1.15 | 8,336 |
| F: Commute trip duration* | 0 | 165 | 15.37 | 16.52 | 1.86 | 5.79 | 4,163 |
| M: Commute trip duration* | 0 | 212 | 23.68 | 20.49 | 2.07 | 8.22 | 4,156 |
| F: Commute trip car use* | 0 | 1 | 0.44 | 0.46 | 0.24 | –1.84 | 4,163 |
| M: Commute trip car use* | 0 | 1 | 0.63 | 0.44 | –0.55 | –1.53 | 4,156 |
| Urbanity at residence** | 0 | 5 | 3.12 | 1.44 | –0.47 | –0.61 | 8,364 |
| Year of survey | 1994 | 2008 | 2001.81 | 3.97 | –0.13 | –1.07 | 8,364 |
| Traditionalism of houschold in employment | 0 | 1 | 0.67 | 0.31 | –0.77 | –0.28 | 4,341 |
| Traditionalism of household in household work | 0 | 1 | 0.62 | 0.25 | –0.28 | –0.57 | 4,568 |

F: female, M: male.

* Commute trip duration is measured as the mean duration of job and education trips over the report week. Commute trip car use is measured as the proportion of job/education trips made driving a car among all job/education trips. Respondents without job or education trips are coded as zero as commute time constraints do not apply.

** Urbanity is measured as the diversity of activity opportunities in the neighbourhood. The questionnaire includes a range of dummy variables indicating whether or not certain facilities are accessible on foot in the neighbourhood (grocery shopping, other shopping, pubs/restaurants, cultural facilities, sports facilities). The number of choices are combined in a sum score (Cronbach's $\alpha$=0.73).

As this research focuses on the two partners' relative contributions to the workload, relative measures of the two partners' education, employment and age category are used. For instance, I hypothesize that if a woman has a higher level of education than her partner, she will have more career expectations than her partner. Accordingly, her partner would participate more than average in household chores. 'Dominating strongly' in education means that the respondent's achievement is at least two degrees higher than his/her partner's degree.

Concerning age categories, I assume that the elder partner tends to have more authority. In households where the female partner is the elder one, this should result in less traditionalism, as her negotiating position is relatively good.

Additionally, level variables are used. For example, the household education level is represented by the highest educational qualifications from either partner. The number of children is included in three age brackets, assuming that all are associated positively with traditionalism, especially infants.

The inclusion of employment, mean duration of job and/or education trips and car use for job/education trips are all limited to the household work model in order to avoid tautological explanations. One may expect more responsibility for household chores among respondents with short commutes, as well as among respondents who use the car and are thus more flexible (Schwanen 2007). On the other hand, households owning as many cars as drivers are hypothesized to organize their housework more traditionally. This is due to the permanent availability of the car to the female partner, which may mean that she would not commit her partner to participate as much in household chores. Despite this, I expect car availability to be associated with less traditionalism in employment, since a high level of car availability should improve women's chances for integration into the labour market.

Concerning spatial context in the city (and in East Germany), more egalitarian divisions of labour are hypothesized to commonly exist. Due to segregation, which is particularly prevalent in cities, two variables represent the neighbourhood in addition to municipality size: central vs. remote location and urbanity.

As in most travel surveys, the data used do not include information on personal income. Household income is included only from 2001. It is transformed into per-capita-income weighted by age category of the household members (OECD 'equivalent income' scale) and used only for control analyses.

(4) The data also do not include direct information on household type. Household type must therefore be reconstructed from the number of household members along with their sex and age. Couples with or without children are both considered here.

I define those households as families that have one or more children below the age of 18, as well as households with three or more adults if the age difference between the second and the third oldest household member is more than 18 years, and the third oldest household member is younger than 30 years of age ('family with adult child(ren)'). Households with three or more adults that do not match both these conditions are classified as 'other households' and therefore excluded from analysis.[4]

In the case of complex households, this procedure does not allow for a clear decision regarding the household members that are assumed to be partners. These cases were inspected individually. Homosexual couples have been excluded from analysis (0.5 per cent of all couples and families).

## Results

### *Time expenditure for activities and trips*

As expected, men dominate employment (job, business), while women dominate household work (shopping/errands, escort) (Table 4.3). This refers to both activities and trips. In business, gender differences are particularly pronounced and women account for a mere 22 per cent of the total time spent by couples. This distinct difference suggests that women tend to work as dependent employees rather than being self-employed, and even self-employed women make fewer business trips than their male counterparts.

Concerning household work, women account for 56 per cent of a household's total time expenditure for shopping/errands and 58 per cent of total time expenditure for escorting other persons. Again these figures are similar, but less pronounced, for trip duration. Hence, gender inequality seems to be more prevalent in employment than in household work (domestic work unconsidered).

Education, leisure, housing and other activities exhibit minor gender differences. Women's somewhat longer in-home time expenditure meets the expectation that there is less mobility among women. It may also reflect more time spent for domestic chores.

These results vary strongly by household type. Children under 14 years of age are relatively dependent on parental care and escort. In such households, the distribution of employment and household chores is strongly gendered (Table 4.4). Women's share in both partners' employed work is only 24 per cent while their share in household work is 65 per cent and, more specifically, even 71 per cent in escort trips. This means that men spend more than three times as much time on employment than women, while women spend almost twice as much time as men on household work (excluding domestic work).

4 False coding may occur anyway. Imagine a patchwork family composed of a 49-year-old man, his 25-year-old second wife, the 24-year-old husband's son and the 3-year-old wife's daughter. This household would be classified as a young family with a grandfather living in the household. The 25-year-old woman and the 24-year-old man would be classified as a couple. Cases such as this are likely to be very rare, however. Also note that all those identified as couples may be flat-mates rather than couples in reality.

**Table 4.3 Activity and trip time expenditure in couples by gender**

| | Job | Business | Shopping/ errands | Escort | Education | Leisure | Other | All out-of-home activities | At home*** | n |
|---|---|---|---|---|---|---|---|---|---|---|
| | Activity time expenditure (min per week) | | | | | | | | | |
| Men | 1,192* | 173* | 185* | 34* | 17 | 1,103 | 11 | 2,715* | 5,464* | 6,296 |
| Women | 663* | 49* | 235* | 47* | 20 | 1,063 | 9 | 2,086* | 5,866* | 6,296 |
| Women (%)** | 35.7 | 22.2 | 55.9 | 58.1 | 53.3 | 49.1 | 45.9 | 43.4 | 51.8 | |
| | Trip time expenditure (min per week) | | | | | | | | | |
| Men | 133* | 69* | 112* | 29* | 4 | 262(*) | 7 | 616* | – | 6,296 |
| Women | 75* | 19* | 128* | 37* | 5 | 243(*) | 6 | 512* | – | 6,296 |
| Women (%)** | 35.9 | 21.6 | 53.2 | 56.0 | 51.0 | 48.2 | 45.9 | 45.4 | | |

Partners in couples and families with information for both partners, excluding homosexual couples, 1994–2008. Note that the analysis is based on pooled data. Control analyses with a sample of independent observations (including only one record per household) yield extremely similar results.

* significant (p=0.01)

(*) significant (p=0.05)

** woman's share of total expenditure of both partners

*** excluding days without trip records (holidays, sickness, 'abnormal' days, missing)

**Table 4.4 Time expenditure (minutes per week, sum of activities and trips) in couples by gender and household type**

| | Employment | | | Household work | | | |
|---|---|---|---|---|---|---|---|
| | Job | Business | Total | Shopping/ errands | Escort | Total | n |
| **Households without children < 14 years** | | | | | | | |
| Men | 987* | 214* | 1201* | 336* | 60* | 396* | 4,298 |
| Women | 764* | 63* | 828* | 369* | 43* | 413* | 4,298 |
| Women (%)** | 43.6 | 22.9 | 40.8 | 52.4 | 42.0 | 51.0 | |
| **Households with children < 14 years** | | | | | | | |
| Men | 2,087* | 304* | 2,391* | 209* | 71* | 281* | 1,998 |
| Women | 679* | 79* | 759* | 347* | 172* | 520* | 1,998 |
| Women (%)** | 24.6 | 20.6 | 24.1 | 62.4 | 70.8 | 64.9 | |

See remarks below Table 4.3. Tables categorized by activities and trips separately are available from the author.

**Table 4.5 Time expenditure (minutes per week, sum of activities and trips) in couples by gender and household type**

| | 1994–96 | 1997–99 | 2000–02 | 2003–05 | 2006–08 |
|---|---|---|---|---|---|
| **Households without children < 14 years** | | | | | |
| Employment | | | | | |
| Men | 1,247* | 1,248* | 1,171* | 1,185* | 1,159* |
| Women | 860* | 787* | 759* | 851* | 891* |
| Women (%)** | 40.8 | 38.7 | 39.3 | 41.8 | 43.5 |
| Household work | | | | | |
| Men | 363 | 393* | 399(*) | 418 | 402 |
| Women | 373 | 429* | 417(*) | 430 | 409 |
| Women (%)** | 50.7 | 52.2 | 51.1 | 50.7 | 50.4 |
| **Households with children < 14 years** | | | | | |
| Employment | | | | | |
| Men | 2,421* | 2,353* | 2,423* | 2,357* | 2,401* |
| Women | 584* | 693* | 839* | 769* | 959* |
| Women (%)** | 19.4 | 22.7 | 25.7 | 24.6 | 28.5 |
| Household work | | | | | |
| Men | 249* | 321* | 268* | 296* | 272* |
| Women | 532* | 523* | 484* | 540* | 517* |
| Women (%)** | 68.1 | 62.0 | 64.3 | 64.6 | 65.5 |
| n (Households without children) | 547 | 999 | 987 | 961 | 804 |
| n (Households with children) | 310 | 453 | 418 | 425 | 392 |

Test of significance: gender comparison within study year category.

See more remarks below Table 4.3.

In couples either without children or with older children, the situation is clearly different. Both household and employed work is distributed to a large extent more equally. Despite this though, pronounced gender differences can be found once again with regard to business. These findings are true for retirees as well as for younger couples without children (without table).

Table 4.5 shows trends during the study period. The evidence for changes over time in the division of work is limited. No clear trends can be detected at all in households without children under 14. In households with young children, women's share in employed work clearly increases. A regression model performed analogously to those presented below estimates the increase in women's share in employment to be 0.4 per cent per annum (not documented here). An inspection of absolute values suggests that this trend is caused by an increase in female employment, while men's values are relatively constant. This trend is not compensated for in the realm of household work though. Still, households with younger children clearly have more traditional work-sharing arrangements.

*Work-sharing in partnerships – household types*

The mean value comparisons presented above do not allow for a conclusion on how many households exercise which type of work-sharing. Therefore, the potential combinations of employment and household work are categorized in the nine types of work-sharing shown in Table 4.6.

**Table 4.6 Types of intra-household employment and household work distribution in couples and family households**

| Type | | All households | Children < 14 years | |
|---|---|---|---|---|
| | | | no* | yes* |
| 1 | Traditional (M dominates employment, F dominates household) | 40.2 | 27.1 | 59.4 |
| 2 | Reversed roles (F dominates employment, M dominates household) | 8.0 | 11.0 | 3.5 |
| 3 | M dominates employment and household | 6.8 | 7.2 | 6.3 |
| 4 | F dominates employment and household | 3.6 | 4.3 | 2.5 |
| | Equity … | | | |
| 5 | … in employment and household | 7.4 | 9.9 | 3.6 |
| 6 | … in employment, F dominates household | 8.9 | 11.4 | 5.3 |
| 7 | … in employment, M dominates household | 5.7 | 7.8 | 2.7 |

| Type | | All households | Children < 14 years | |
|---|---|---|---|---|
| | | | no* | yes* |
| | Equity … | | | |
| 8 | … in household, F dominates employment | 4.9 | 6.5 | 2.5 |
| 9 | … in household, M dominates employment | 14.5 | 14.7 | 14.2 |
| | Total | 100.0 | 100.0 | 100.0 |
| | n | 4,570 | 2,762 | 1,808 |

* Significant difference (p=0.01)

F: female; M: male partner. 1994–2008.

About four out of ten couples practise a traditional division of labour, where the male partner dominates employment while the female partner dominates household work. Eight per cent of the households practise a reversed roles model. In 7 per cent of the households the male partner undertakes the larger share in both employment and household work, while in just under 4 per cent the female partner does so. The remaining 41 per cent of households exercise relative equality in at least one of the two realms of work. However, in a few households (7 per cent), the partners contribute equally to both employment and household work. Other couples share employment equally between each other, but do not share household work equally or vice versa.

In 27 per cent of the households the male partner generally dominates out-of-home work (types 3, 7, and 9), or in other words, he dominates either both realms of work or at least one of the two realms with the other one shared equally. In 17 per cent of the households the female partner is generally dominant (types 4, 6, and 8). Due to the higher frequency of full-time employment among men, this inequality appears considerably more strongly as soon as the sum of household and employed work is considered (Table 4.7). In almost half of the households (48 per cent) the male dominates the female time expenditure for out-of-home work. In 36 per cent of the households the distribution is relatively equal.

**Table 4.7 Distribution of intra-household work time expenditure in couples and family households**

| | Sum of employment and household work |
|---|---|
| M dominates | 48.0 |
| F dominates | 16.3 |
| Equity | 35.8 |
| Sum | 100.0 |
| n | 6,162 |

See more remarks below Table 4.6.

It is important to note that these distributions vary strongly by household type. A striking finding is that the share of households exercising a traditional division of work is more than twice as large in households with young children as in other households (Table 4.6). The reversed roles model is practised in not more than 3 per cent of households with children, but is practiced three to four times more frequently (11 per cent) in other households. Other types of relatively equal arrangements (types 5 to 9) are considerably less frequent in households with children as well, with the exception of type 9. This traditionalizing effect of children appears to become weaker as children grow up though. In 62 per cent of households with children under 10, the traditional arrangement is exercised (table available from the author). The respective figure is 'only' 47 per cent in households where the youngest child is between the ages of 10 and 13, and 38 per cent in households where the youngest child is between the ages of 14 and 17. In households with adult children the figure is 28 per cent and in households without children it is 25 per cent. Although cohort effects may play a role, traditionalism among the youngest parents' cohort does not seem plausible to the extent the figures suggest. It thus seems likely that the traditionalism emerging after the foundation of a family is reduced as the children grow older.

Is inequality more pronounced in employment or in household work? By considering the two realms of work separately, about one fifth of households exercise a strongly traditional division of household work (Table 4.9), which means that the female partner undertakes more than 80 per cent of the household work. However, in even more households the roles are reversed (23–4 per cent). One third of the couples share their household work equally. Turning to employment work-sharing, considerably more households (four out of ten) practise a strongly traditional model (Table 4.10). Reversed roles as well as egalitarian patterns are less frequent in employment than in housework. Therefore, there seems to be more traditionalism in employment than in housework.

Notable changes exist throughout the study period (Table 4.8). Traditional patterns tend to decline while reversed roles become more common. Starting from a low level, other unconventional patterns, such as egalitarian household work-sharing combined with female dominance in employment, increase. What is more, the share of households in which the female partner dominates both employment and household work, increases. The egalitarian distribution of both realms of work shows no clear trend.

Are these changes due to the redistribution of employment or of household work? Overall, the changes in employment are considerably stronger than those in household work (Table 4.9, Table 4.10). One may argue that the existence of a more traditional work-sharing in employment provides more potential for change towards equity, but this change hardly leads to more equity. Instead, reversed role arrangements increase over time. At the aggregate level (but not at the household level) this leads to more gender equity.

**Table 4.8 Trends in intra-household employment and household work distribution in couples and family households**

| Type | | 1994–96 | 1997–99 | 2000–02 | 2003–05 | 2006–08 |
|---|---|---|---|---|---|---|
| 1 | Traditional (M dominates employment, F dominates household) | 44.6 | 42.5 | 40.6 | 37.2 | 35.1 |
| 2 | Reversed roles (F dominates employment, M dominates household) | 6.6 | 7.6 | 8.9 | 7.7 | 9.3 |
| 3 | M dominates employment and household | 6.8 | 8.0 | 5.7 | 6.8 | 6.9 |
| 4 | F dominates employment and household | 2.9 | 3.2 | 3.8 | 3.9 | 4.3 |
| | Equity … | | | | | |
| 5 | … in employment and household | 6.0 | 9.1 | 7.2 | 7.0 | 7.6 |
| 6 | … in employment, F dominates household | 8.9 | 7.9 | 8.3 | 10.2 | 9.4 |
| 7 | … in employment, M dominates household | 5.8 | 4.2 | 5.3 | 6.9 | 6.5 |
| 8 | … in household, F dominates employment | 3.4 | 3.5 | 5.4 | 5.8 | 6.7 |
| 9 | … in household, M dominates employment | 15.0 | 14.0 | 14.6 | 14.5 | 14.3 |
| | Total | 100.0 | 100.0 | 100.0 | 100.0 | 100.0 |
| | n | 671 | 1,071 | 979 | 1,004 | 845 |

Differences between time brackets significant (p=0.05)

**Table 4.9 Trends in household work distribution in couples and family households**

| | 1994–96 | 1997–99 | 2000–02 | 2003–05 | 2006–08 |
|---|---|---|---|---|---|
| Strongly traditional | 21.8 | 20.2 | 19.9 | 19.6 | 17.6 |
| Traditional | 24.4 | 25.8 | 24.5 | 23.5 | 24.5 |
| Reversed roles | 23.0 | 23.2 | 23.2 | 24.1 | 24.3 |
| Egalitarian | 30.8 | 30.8 | 32.5 | 32.8 | 33.6 |
| Total | 100.0 | 100.0 | 100.0 | 100.0 | 100.0 |
| n | 849 | 1,426 | 1,359 | 1,361 | 1,153 |

Differences between time brackets not significant (p=0.05)

**Table 4.10 Trends in employed work distribution in couples and family households**

| | 1994–96 | 1997–99 | 2000–02 | 2003–05 | 2006–08 |
|---|---|---|---|---|---|
| Strongly traditional | 43.7 | 45.1 | 40.0 | 37.8 | 36.1 |
| Traditional | 22.6 | 19.2 | 20.9 | 20.7 | 20.0 |
| Reversed roles | 12.8 | 14.3 | 18.2 | 17.3 | 20.3 |
| Egalitarian | 20.9 | 21.5 | 20.9 | 24.2 | 23.6 |
| Total | 100.0 | 100.0 | 100.0 | 100.0 | 100.0 |
| n | 672 | 1,075 | 983 | 1,008 | 849 |

Differences between time brackets significant ($p=0.01$)

The trend towards reversed roles (rather than equal distributions) means that at a minimum, opportunity equality should increase over time. An increase in reversed role constellations may as well imply more room for negotiation in partnerships, even if there is no trend towards more equity in outcome.

Traditional arrangements tend to decline in household work-sharing too. This decline tends to favour egalitarian roles rather than reversed ones. These changes are less notable than in employment though, and how they are judged depends on the perspective taken. The observed decline of strongly traditional role models in household work-sharing measures to one percentage point in three years; this trend may be less than striking, but it is clearly more than nothing provided it continues in the future.

*Impact factors of within-household work-sharing*

Which factors may help explain within-household distributions of work? Table 4.11 shows two regression models of traditionalism in work-sharing arrangements. Multicollinearity issues were satisfactory (all variance inflation factors <5). Effect strengths are represented by standardized coefficients Beta in the table.

Employment shows the strongest effect on **household work-sharing**. Education level, household type, car availability and commuting behaviour are significant factors as well. Effects of age group are minor. Spatial context has no significant impact (interpretation of significance refers to $p<0.05$).

As far as employment and education are concerned, the relative level of two partners (as compared to one another) as well as the total household level is important. Strong male dominance in employment tends to be associated with a strongly traditional division of household work. In cases where the male partner has a weaker dominance in employment, the effect is more moderate. Female partners' dominance in employment is associated with less traditional household work-sharing arrangements.

**Table 4.11 Impact factors of traditionalism in household work and employment**

| | Traditionalism (household work) | | | Traditionalism (employed work) | | |
|---|---|---|---|---|---|---|
| | **B** | **Beta** | **Sig.** | **B** | **Beta** | **Sig.** |
| Constant | 0.17 | | 0.93 | 5.40 | | 0.02 |
| Age category (Ref.: both partners 18–39 years) | | | | | | |
| M 18–39, F 40–59 years | 0.04 | 0.02 | 0.11 | –0.04 | –0.02 | 0.26 |
| M 40–59, F 18–39 years | 0.01 | 0.01 | 0.41 | 0.00 | 0.00 | 0.85 |
| M 40–59, F 40–59 years | 0.01 | 0.03 | 0.19 | 0.01 | 0.01 | 0.65 |
| M < 60, F 60+ years | –0.08 | –0.04 | 0.02 | 0.23 | 0.08 | 0.00 |
| M 60+, F < 60 years | –0.03 | –0.03 | 0.14 | –0.19 | –0.15 | 0.00 |
| M 60+, F 60+ years | –0.03 | –0.03 | 0.13 | 0.01 | 0.00 | 0.82 |
| No. of children < 10 years in household | 0.03 | 0.07 | 0.00 | 0.12 | 0.27 | 0.00 |
| No. of children 10–13 years in household | 0.01 | 0.02 | 0.29 | 0.04 | 0.05 | 0.00 |
| No. of children 14–17 years in household | 0.00 | 0.00 | 0.98 | 0.01 | 0.02 | 0.27 |
| Relative education level (Ref.: both partners have same education level) | | | | | | |
| M: dominates strongly | –0.02 | –0.01 | 0.45 | 0.09 | 0.07 | 0.00 |
| M: dominates somewhat | 0.04 | 0.06 | 0.03 | 0.03 | 0.03 | 0.20 |
| F: dominates somewhat | 0.09 | 0.14 | 0.00 | –0.01 | –0.02 | 0.54 |
| F: dominates strongly | 0.04 | 0.03 | 0.11 | –0.04 | –0.02 | 0.16 |
| Education level of household (Ref.: both partners have no degree or elementary school) At least 1 partner has … | | | | | | |
| … elementary school degree with apprenticeship | –0.04 | –0.08 | 0.03 | 0.00 | 0.00 | 0.93 |
| … secondary school level I | –0.05 | –0.09 | 0.01 | 0.01 | 0.02 | 0.53 |
| … final secondary school degree (university entrance) or university degree | –0.03 | –0.06 | 0.08 | 0.02 | 0.03 | 0.32 |
| Relative employment level (Ref.: both partners employed to the same extent) | | | | | | |
| M: dominates strongly | 0.19 | 0.33 | 0.00 | | | |
| M: dominates somewhat | 0.10 | 0.20 | 0.00 | | | |
| F: dominates somewhat | –0.13 | –0.14 | 0.00 | | | |
| F: dominates strongly | –0.10 | –0.08 | 0.00 | | | |
| Employment level of household: at least one partner is full-time employed | –0.06 | –0.09 | 0.00 | | | |

| | | | | | | |
|---|---|---|---|---|---|---|
| Car availability (number of cars minus number of licensed household members) | 0.03 | 0.09 | 0.00 | 0.03 | 0.06 | 0.00 |
| F: Commute trip duration | 0.00 | –0.04 | 0.01 | | | |
| M: Commute trip duration | 0.00 | 0.04 | 0.01 | | | |
| F: Commute trip car use | 0.07 | 0.13 | 0.00 | | | |
| M: Commute trip car use | –0.03 | –0.04 | 0.01 | | | |
| Municipality size category (Ref.: < 20.000 inh) | | | | | | |
| 20–100.000 inh | 0.00 | 0.00 | 0.89 | –0.01 | –0.01 | 0.50 |
| 100–500.000 inh | –0.02 | –0.02 | 0.18 | –0.03 | –0.03 | 0.04 |
| > 500.000 inh | 0.00 | 0.00 | 0.83 | –0.04 | –0.05 | 0.00 |
| Central residential location within municipality | 0.00 | –0.01 | 0.68 | 0.03 | 0.04 | 0.02 |
| Urbanity at residence | 0.00 | 0.00 | 0.85 | 0.00 | –0.01 | 0.69 |
| East Germany | 0.01 | 0.01 | 0.35 | –0.08 | –0.10 | 0.00 |
| Year of survey | 0.00 | 0.00 | 0.83 | 0.00 | –0.03 | 0.05 |
| $R^2$ (adj) | 18.0 | | | 14.4 | | |
| n | 3,564 | | | 4,075 | | |

M: male, F: female.

The absolute level of household employment is significant as well. The full-time employment of at least one partner is associated with less traditionalism. This could be an effect of higher social status levels, which may reduce traditional arrangements. This interpretation is supported by the observation that a high household education level also tends to be associated with less traditionalism, although the effect of the highest level is not significant.

Turning to relative education levels, one may expect that traditionalism increases as the relative status of the male partner increases. This is, in part, due to the fact that male dominance in education level provides increased opportunities to realize a high income level. From this logic, a high relative education level of the female partner should work in the opposite direction. However, this is not reflected in the data.

By contrast, female dominance in education status tends to be associated with *more* traditionalism when compared to the reference category. The same is true for male dominance, albeit to a lesser extent. This surprising finding contradicts the assumption that highly educated individuals tend to have an innovative role in modernization. The result is possibly less a reflection of traditionalism in ideologies than a reflection of a relatively large amount of time spent on shopping and escort activities, for which women with a lower level of education may not (to the same extent) have the required resources, such as time, money and cars. Another interpretation could be that male partners with a low relative education level compared to their wife tend to demonstrate masculinity by engaging in employment rather than in housework.

The effects children have in the household confirm previous expectations as well as the descriptive findings above. This is especially true for children under 10 years of age who typically require the most care. Clearly, young children are associated with more traditional household work-sharing while older children have no significant effect.

A high level of car availability is associated with more traditionalism as well. It should be kept in mind that this relationship is not necessarily unidirectional though. A traditional role model may fuel a household's desire to purchase at least a second car in order to alleviate the wife's household work. The reversed causal relationship is suggested here, or in other words the use of the car is likely to encourage traditionalism, as its use facilitates goods transport and the coupling of activities. Thus, car use enhances women's abilities to undertake additional trips; for instance, to escort children to school or to do the shopping after work. Hence, the use of the car may impair the negotiating position of females with their partners concerning household work. In any case, not only the availability, but also the use of a car by the female partner for her work trips corresponds markedly with more traditionalism.

For males, the effect of the car is rather the opposite. The male partner's use of the car for commuting is associated with less traditionalism. The relatively weak effect magnitude for men suggests that car use may impair the negotiating position of men concerning household duties to a lesser extent than it impairs women. In other words, a man's negotiating power appears to be stronger than that of a woman.

Commute trip duration decreases the time budget for other activities and acts similarly to car use. Long commute trips by women decrease traditionalism and strengthen the female negotiating position, while long commute trips by men increase traditionalism. The magnitude of effects does not suggest any patriarchal structures here, as opposed to car use.

In the traditionalism model for **employed work** the strongest effects can be seen in specific age categories and household types. Children under 10 years of age boost traditionalism in employment even more than in household work-sharing, although this picture is likely to look different for in-home household duties. Children between the ages of 10 and 13 also have this effect, although to a considerably lesser extent. Descriptive analysis shows this is mainly due to more employed work time expenditure among women with elder children, compared to women with infants. In contrast, men's time expenditure for employment decreases slightly with the age of the youngest child. One may speculate that men increase their employed work to compensate for the women's lack of income while women care for young children (or they use employment in order to withdraw from education duties).

The effects of age are gender specific. Compared with the reference group of both partners between the ages of 18 and 39, employment work-sharing is more traditional when the female partner is 60 years or older and her partner is younger as well as vice versa. This refers to households in which one of the partners is already retired or at least in part-time employment prior to retirement.

The effects of a partner's education level are weaker here than in the household work model and they are not significant. The effects of relative education are plausible. Here, analysis suggests that work-sharing arrangements are relatively traditional for couples with a strong male dominance in education. On the other hand, a higher female education level is associated with less traditionalism (as compared to equity in education level).

Car availability again has a traditionalizing impact, but is less obvious in employed work than in household work. Here, one might expect a negative association rather than a positive one. A household with a higher income is associated with more cars, and with more cars, a woman's access to the labour market is improved. It is possible that sociocultural backgrounds such as egalitarian work-sharing arrangements (low traditionalism) may be at work here. They may also be associated with car-critical attitudes and, accordingly, a low level of car availability.

As opposed to the household work model, the effects of residential spatial context are significant here. In large cities traditionalism levels are lower than in small municipalities. Contrary to expectations, traditionalism is higher in central locations within cities than at the urban fringe. In East Germany, traditionalism in employment work-sharing is considerably lower than in West Germany, confirming previous knowledge (Gille and Marbach 2004). The year of survey has a significantly negative effect; in other words, traditionalism is declining.

As noted above, the use of income in the models was limited to control analyses, as it was only recorded as of 2001. According to these control analyses, the results are largely stable. There were no cases of inverted signs in significant effects. Equivalent income was not significant in the household work model. In the employment model, higher income was associated with less traditionalism, confirming expectations.

## Conclusions

In this chapter, time-use data for trips and out-of-home activities have been used to study the extent of traditionalism in couples and family households. The results confirm established knowledge on differences in time expenditure between partners. Men spend more time on employed work than women, while women spend more time on household work. The gender differences in household work are less pronounced. The gender roles are particularly traditional in households with children.

However, traditionalism has decreased during the study period of 1994 to 2008, particularly in families. This refers mainly to employment, while the gendered distribution of household work remained persistently unequal. Increasing employment among mothers is thus not counterbalanced by a decrease in household duties. This is not consistent with the German time use survey, in which a converging gender participation in household work was found for the period of 1991/92 to 2001/02 (Gille and Marbach 2004). These inconsistencies may be due to my focus on out-of-home work. However, the findings in this chapter confirm the observation of the 'persistence of employed wives' primary responsibility for domestic labor' (Bianchi et al. 2000: 197–8) made elsewhere.

About 40 per cent of couples and family households may be assigned to a traditional role model combination. This figure sounds perhaps surprisingly low, particularly as almost half of German households are single or single-parent (equivalent to about 30 per cent of the population) for whom this model is in any case not valid. It has to be kept in mind, however, that this chapter did not consider domestic work. The other 60 per cent of households exercise various patterns of work-sharing, such as egalitarian or reversed roles, or the dominance of the male or female partner in employment as well as household work. Concerning household work, egalitarian or reversed role models are more common than in employment. However, gender patterns in employment show considerably more change over time.

The extent of traditionalism in a household is affected by various conditions including household type, relative social status (in terms of education and employment), time budgets and commute travel patterns. Some spatial context effects are also significant, but these appear to be minor.

Overall, predominant impact factors on gender equality can be found. First, the presence of children in the household strongly corresponds with traditional work-sharing arrangements between their parents. This is especially true when children are still young. Second, the relative employment level of each of the partners strongly impacts gender roles. Here, male dominance in employed work is associated with a marked traditionalism in houschold work-sharing, again confirming established knowledge and suggesting both a household related rationale in work-sharing and male economic dominance. Such associations clearly reflect interdependencies between two partners' activity patterns, which can be interpreted as time-geographical authority and coupling constraints.

Travel behaviour also strongly affects the division of work. Female use of the car for commuting strengthens their commitment to shopping and escort trips. Note that this does not necessarily reflect a unidirectional causal relationship, since a woman's commitment may actually enforce her use of the car. The magnitude of the effect, compared to male car use, suggests that women have a relatively weak negotiating position when it comes to housework duties. However, this is not confirmed by gender specific effect magnitudes in commute duration.

The overall analysis suggests that economic and temporal resources as well as the given context affect the division of labour within households. Economic inequalities on the labour market are themselves affected by gender specific responsibilities, social roles and perhaps even preferences, which may be reflected in the strong effects exhibited by the presence of infants on work-sharing arrangements.

Besides contributing to understanding intra-household micro-structures, these results also shed light on sustainability issues in terms of gender equity in travel and activity patterns. Some conclusions may be drawn for further research.

First, the balance of power between preferences/attitudes and unequal resources is still a largely open question. This is reflected, for instance, in the ambivalent interpretations of the observation that women work for below-average income. Most feminist theorists would argue that patriarchal power relations are at work here while others assign this observation to women's lack of *will* to assert their interests (Babcock and Laschever 2003).

Second, the complex relationship between preferences and resources – 'freedom and constraint' – that are themselves both subject to (unsustainable) gender inequalities, calls for more complex approaches to analysis, such as structural equation modelling or Bayesian networks. These could also possibly be complemented by qualitative approaches. All these approaches can address interactions between various activities in addition to employment and housework, especially between in-home and out-of-home activities. The majority of sociological studies in this field focus on housework, and resources such as the private car or commute duration rarely play a role in such studies. On the other hand, related studies in the transport field, such as the present one, are almost exclusively focused on out-of-home activities. Further integrating related disciplines – for which there is already much evidence – represents a very valuable future direction from a theoretical perspective as well as for empirical investigation.

## Acknowledgement

This research was funded by the German Research Foundation (DFG) as part of the project 'Alltag im Wandel des Geschlechterverhältnisses: Aktivitäten, Wege, Verkehrsmittel und Zeitverwendung' ('Everyday life in the context of changing gender relations: activities, trips, travel modes and time use', 2009–2013).

## References

Babcock, L. and Laschever, S. 2003. *Women Don't Ask: Negotiation and the Gender Divide*. Princeton: Princeton University Press.

Baxter, J. 1997. Gender equality and participation in housework: a cross-national perspective. *Journal of Comparative Family Studies*, 28 (3), 220–47.

Beck, U. 1992. *Risk Society*. New Delhi: Sage.

Bianchi, S.M., Milkie, M.A., Sayer, L.C. and Robinson, J.P. 2000. Is anyone doing the housework? Trends in the gender division of household labor. *Social Forces*, 79 (1), 191–228.

Blau, F.D., Ferber, M.A. and Winkler, A.E. 2010. *The Economics of Women, Men, and Work*. 6th edition. Englewood Cliffs, NJ: Prentice-Hall.

BMFSFJ and Statistisches Bundesamt (ed.) 2003. *Wo bleibt die Zeit? Die Zeitverwendung der Bevölkerung in Deutschland 2001/02*. Wiesbaden: Statistisches Bundesamt.

Chlond, B. and Kuhnimhof, T. 2005. *Panelhandbuch. Das Deutsche Mobilitätspanel: Informationen sowie Anleitung und Anregungen zur Datennutzung*. Karlsruhe: Universität.

Cooke, L.P. 2006. Policy, preferences, and patriarchy: the division of domestic labor in East Germany, West Germany, and the United States. *Social Politics*, 13 (1), 117–43.

Ettema, D. and Van der Lippe, T. 2009. Weekly rhythms in task and time allocation of households. *Transportation*, 36 (2), 113–29.

Gille, M. and Marbach, J. 2004. Arbeitsteilung von Paaren und ihre Belastung mit Zeitstress, in *Alltag in Deutschland. Analysen zur Zeitverwendung*, edited by Statistisches Bundesamt. Wiesbaden: Statistisches Bundesamt, 86–113.

Grunow, D., Schulz, F. and Blossfeld, H.-P. 2007. Was erklärt die Traditionalisierungsprozesse häuslicher Arbeitsteilung im Eheverlauf: soziale Normen oder ökonomische Ressourcen? *Zeitschrift für Soziologie*, 36 (3), 162–81.

Hägerstrand, T. 1970. What about people in regional science? *Papers in Regional Science*, 24 (1), 7–21.

Hanson, S. 2010. Gender and mobility: new approaches for informing sustainability. *Gender, Place and Culture*, 17 (1), 5–23.

Hanson, S. and Pratt, G. 1988. Spatial dimensions of the gender division of labor in a local labor market. *Urban Geography*, 9 (2), 180–202.

Law, R. 1999. Beyond 'women and transport': towards new geographies of gender and daily mobility. *Progress in Human Geography*, 23 (4), 567–88.

McGuckin, N. and Murakami, E. 1999. Examining trip-chaining behavior: a comparison of travel by men and women. *Transportation Research Record*, 1693, 79–85.

Marks, J.L., Lam, C.B. and McHale, S.M. 2009. Family patterns of gender role attitudes. *Sex Roles*, 61 (3–4), 221–34.

Root, A., Schintler, L. and Button, K. 2000. Women, travel and the idea of 'sustainable transport'. *Transport Reviews*, 20 (3), 369–83.

Scheiner, J. and Holz-Rau, C. 2012. Gender structures in car availability in car deficient households. *Research in Transportation Economics*, 34 (1), 16–26.

Schwanen, T. 2007. Gender differences in chauffeuring children among dual-earner families. *The Professional Geographer*, 59 (4), 447–62.

Statistisches Bundesamt (ed.) 2004. *Alltag in Deutschland. Analysen zur Zeitverwendung*. Wiesbaden: Statistisches Bundesamt.

Uteng, T.P. and Cresswell, T. 2008. *Gendered Mobilities*. Aldershot: Ashgate.

Wagner, M. and Franzmann, G. 2000. Die Pluralisierung der Lebensformen. *Zeitschrift für Bevölkerungswissenschaft*, 25 (1), 151–73.

# PART II
# Models for Sustainability

Chapter 5

# Air Pollution Hotspots in Urban Areas – How Effective are Pricing Strategies to Comply with the EU Limits for $NO_2$?

Friederike Hülsmann, Benjamin Kickhöfer and Regine Gerike

## Introduction

When discussing sustainable mobility, one core element is the conservation of natural resources to keep the environment liveable for present and future generations. Packages of measures are needed for achieving this ambitious goal (Gehlert et al. 2011, Gerike et al. 2008). These packages combine highly effective measures of low public acceptance, such as pricing measures and driving bans for trucks, with less effective measures of high acceptance, such as information and improvements in the transport supply of environmentally friendly modes. Using this approach, significant reductions in environmental impacts can be achieved while at the same time gaining public acceptance, a vital pre-condition for the successful implementation of any strategy.

Air pollutant concentration is one of the most important environmental effects of transportation. This topic is constantly under discussion in the European Union (EU) as many cities do not meet the limit values set in the Directive 2008/50/EC on ambient air quality and cleaner air for Europe. High air pollutant concentrations lead to adverse effects on the environment and human health, especially respiratory and cardiovascular diseases (Bickel and Friedrich 2005). This chapter focuses on a pricing measure as a possible strategy for achieving needed reductions in air pollutant concentrations. However, it should be emphasized that a pricing measure can only be implemented if it is embedded in a package of strategies to be sustainably effective. By integrating an 'air quality module' to further develop a policy sensitive approach of transport and emission modelling, this chapter studies the effects of a potential link-based emission toll on local nitrogen dioxide ($NO_2$) concentrations in Munich, Germany. Analysis is conducted for $NO_2$ because it still exceeds the EU limit values in some streets, such as along the street 'Landshuter Allee' (LFU 2012). A toll is introduced on all links that are identified as hotspots in a predefined study area. Starting from an 'emission toll', based on emission cost factors provided by Maibach et al. (2008), this chapter analyses (i) to what extent cost factors have to be increased to remove a significant number of hotspots,

(ii) what the effects on other locations within the study area as well as in the entire city of Munich are and (iii) whether a link-based emission toll is an effective measure to achieve the limit value for $NO_2$ concentrations.

The identification of measurement based hotspots is difficult as only a few measurement stations exist in most cities. Over the past few years, this number has reduced mainly due to economic reasons. In order to decrease $NO_2$ concentrations by designing appropriate policies at relevant locations, all hotspots within a city need to be identified. Determining potential hotspots and concentration levels of $NO_2$ require an integrated approach that considers the complexity of interactions between individuals and the environment. Specifically, it is important to consider the interaction between individual travellers in response to a policy as well as the impact of changes in travel behaviour on $NO_2$ emissions and concentrations, which vary in time and depend on atmospheric conditions and location. As a result, the presented approach is able to identify all relevant hotspots along every street canyon for different policy scenarios.

In the field of traffic-related emission modelling, different approaches have been developed to link travel behaviour to emission calculations. One approach includes macroscopic transport models, which are designed for large scale scenarios and often use MOVES,[1] COPERT[2] or aggregated forms of HBEFA[3] as emission models (EPA 2012, Samaras und Zierock 2009, INFRAS 2010). Microscopic traffic flow simulations are suitable for the analysis of a small study area. Emission calculations are often based on detailed models such as VERSIT+[4] or PHEM[5] (Smit et al. 2006, Zallinger et al. 2010). In Hülsmann et al. (2011) and Kickhöfer et al. (2012, in press), the mesoscopic transport model MATSim is linked to the detailed database of HBEFA. This approach combines both large scale scenarios and detailed emission calculations and is therefore used in this chapter as input for $NO_2$ concentration calculations.

There is an extensive diversity of air quality models that can be applied to local and urban case studies. On the one hand, there are microscopic models, which simulate the dispersion processes in detail (with a high spatial resolution, such as a few street segments), but come with long simulation times. On the other hand, there are urban dispersion models, which simulate dispersion processes for an entire urban area, but come with a low spatial resolution compared to microscopic approaches (Holmes und Morawska 2006). One

1 Motor Vehicle Emission Simulator, see www.epa.gov/otaq/models/moves/index.htm [accessed 15 October 2012].

2 *Computer Program* to calculate Emissions from Road Transport, see www.eea.europa.eu/publications/copert-4-2014-estimating-emissions [accessed 15 October 2012].

3 Handbook on Emission Factors for Road Transport, see www.hbefa.net [accessed 15 October 2012].

4 State-of-the art emission model, see www.tno.nl/downloads/lowres_TNO_VERSIT.pdf [accessed 15 October 2012].

5 Passenger car and Heavy duty Emission Model.

example is the regional and urban dispersion model CALPUFF[6] that was applied by Hatzopoulou and Miller (2010). They used MATSim[7] for the simulation of traffic flows and generated emissions per street segment that were processed in the dispersion model, CALPUFF. The necessary spatial resolution to examine air pollutant concentrations along single street canyons is, however, too low when using CALPUFF. Street canyon modelling is often based on semi-parametric approaches that can be applied locally. Two examples are the CPBM[8] and the OSPM[9]. In contrast to microscopic models, this approach is not able to consider complex building structures and intersections, and simplifies the dispersion processes by empirically determined constants. Instead, street canyon modelling has shorter simulation times, fewer uncertainties and is often well validated. In order to identify the model appropriate for a specific case study, accuracy and availability of the input data need to be fundamental preconditions (Vardoulakis et al. 2003).

The remainder of the present chapter is organized as follows. In the Methodology section, the MATSim simulation framework is presented, which generates the necessary input for modelling air pollutant emissions and concentrations. Emissions are computed individually, based on time-dependent traffic situations and traveller's vehicle attributes. The main features of the air quality module, which uses the output data of the transport simulation and emission modelling tool, are described. This is followed by an explanation of the identification of hotspots and the emission pricing scheme. The Scenario section describes the large-scale real-world scenario of the Munich metropolitan area, the chosen study area for identification of hotspots and for the use of the simulation approach. The goal of decreasing concentrations of $NO_2$ at hotspots as well as the number of hotspots is addressed by applying policy scenarios with a parametric variation of the external emission cost factors. In the Results and Discussion section, the effects of these price variations on $NO_2$ concentration at the hotspots are presented and discussed. Finally, the chapter concludes with a summary of the results and the challenges of the applied approach as well as further research needs.

6 Advanced non steady-state meteorological and air quality modelling system, see http://www.src.com/calpuff/calpuff1.htm [accessed 15 October 2012].

7 Multi-Agent Transport Simulation, see www.matsim.org [accessed 15 October 2012].

8 Canyon Plume Box Model.

9 Operational Street Pollution Model, see http://ospm.dmu.dk/ [accessed 15 October 2012].

## Methodology

### *MATSim*[10]

In the following section, we begin by presenting some brief, general ideas about the transport simulation with MATSim. For a more in-depth discussion of the simulation framework, refer to Raney and Nagel (2006) and the Appendix. In MATSim, each traveller of the real system is modelled as an individual agent. This approach consists of an iterative loop characterized by the following steps:

1. **Plans generation:** All agents independently generate daily plans that encode, among other things, their desired activities during a typical day along with a transport mode for every intervening trip.
2. **Traffic flow simulation:** All selected plans are simultaneously executed in the simulation of the physical system.
3. **Evaluating plans:** All executed plans are evaluated by a utility function which encodes in this chapter the perception of travel time and monetary costs for the available transport modes.
4. **Learning:** Some agents obtain new plans for the next iteration by modifying copies of existing plans. This modification is done by several strategy modules corresponding to the available choice dimensions. The choice between plans is performed within a multinomial logit model.

The repetition of the iteration cycle coupled with the agent database enables the agents to improve their plans over many iterations. This is why it is also called a learning mechanism (see Appendix). The iteration cycle continues until the system has reached a relaxed state. At this point, there is no quantitative measure of when the system is 'relaxed'; we just allow the cycle to continue until the outcome is stable.

### *Exhaust emissions*

The emission modelling tool essentially calculates warm and cold-start emissions for private cars and heavy goods vehicles.[11] The former are emitted when the vehicle's engine is already warmed whereas the latter occurs during the warm-up phase. Warm emissions differ with respect to road type, driving speed, driving dynamics and vehicle characteristics. Cold-start emissions differ with respect to distance travelled, parking time and vehicle characteristics. For the majority of air pollutants, it is found that during cold-start conditions (in comparison to warm engine conditions) more emissions are generated. According to HBEFA, this is

10 Since the methodology remains unaltered, this section is taken from Kickhöfer and Nagel (2012).

11 Public transit (pt) is in the present paper assumed to run emission free.

largely relevant for $NO_2$. It is also important to note that some older gasoline types and more recent diesel-fuel vehicle types exist which generate a lower amount of emissions during the first part of trip. In the scenario, vehicle characteristics are derived from survey data (see the section on Scenario) and include vehicle type, age, cubic capacity and fuel type. Therefore, they can be used for a number of different emission calculations. When no detailed information about the vehicle type is available, fleet averages for Germany are used.

MATSim does not provide detailed driving cycles for each vehicle. Therefore, by taking vehicle-specific free flow and actual travel time for each street segment, the length of a street segment driven with free flow and stop&go can be determined. In the first step of the emission calculation, MATSim driving dynamics are assigned to two traffic situations found in the HBEFA database: free flow and stop&go. For more information on this and the validation of the matching between MATSim and HBEFA driving dynamics, see Hülsmann et al. (2011) and Kickhöfer et al. (2012, in press). HBEFA provides emission factors differentiated for the characteristics presented above. In the second step, so-called 'emission events' are generated and segmented into warm and cold-start emission events. These events provide information about the person, the time, the street segment (= link) and the absolute emitted values by emission type. The definition of emission events follows the MATSim framework, which uses events for storing disaggregated information as objects in JAVA and as XML in output files (see Appendix). For the calculation of air pollutant concentration in the following section, these emission events are summed up per time period and link, resulting in a line emission source with a time resolution of one hour. Emission events are also accessed during the simulation in order to assign cost factors to emissions; thereafter the monetary value of emissions is used for hotspot pricing (see the sub-section on Emission Pricing for 'Hotspots').

*Air pollutant concentration*

This section gives a short introduction to the air quality module, which is developed based on the following criteria:

- Air quality modelling should cover a large scale scenario.
- The spatial resolution of air pollutant concentrations should be directly linked to the spatial resolution of the emissions, the latter of which are link-based.
- Calculation time should be limited to be able to simulate different policy scenarios in a reasonable timeframe.

The Operational Street Pollution Model allows for all three criteria. This street canyon model follows a semi-empirical approach, which is based on a parameterization of the most important dispersion processes close to the street. This includes the influence of the traffic-produced turbulence created by movements of the vehicles, the influence of buildings close to the street on dispersion (street canyon effect) and the chemical transformation between nitrogen monoxide (NO) – ozone ($O_3$) – $NO_2$. OSPM has been successfully tested and applied in many places worldwide (Kakosimos et al. 2010) and recently evaluated in connection with a GIS-based procedure, allowing for calculations in a large number of street segments (Ketzel et al. 2012). The model simulates air pollutant concentration at receptor points, which can be located along any street segment. It should be noted, though, that they are not located very close to an intersection, since dilution processes and additional emissions from the intersecting street cannot be considered. The following data is processed by the air quality module developed for the calculation of $NO_2$ concentration (see also the section on Scenario):

- number of passenger cars and heavy duty vehicles, vehicle speed
- emissions
- meteorology: wind speed, wind direction, ambient temperature, global solar radiation
- background concentration
- street and building geometry
- receptor points

In a simulation step subsequent to the emission modelling, the information on amount and speed of the vehicles, their emissions and the configuration of the street canyon are passed to OSPM. This information is combined with data on wind speed and direction as well as background concentration, in order to determine the pollution concentrations at receptor points at the facade of the buildings. The emissions are aggregated per street cross section because street canyon modelling includes all emissions generated in both directions of a street segment. The OSPM dispersion and chemical processes are then transcribed into an 'OSPM module' written in JAVA to create an integrated tool that combines the generation of the transport activity, the emissions and air pollutant concentration. For nitrogen oxides ($NO_x$), the local contribution from the street is the multiplication of the emission rate with a dispersion parameter resulting from the interactions among wind speed, direction and street canyon geometry. The measurement based urban background concentration is added to determine the total concentration. The total computed $NO_x$ and the measured $O_3$ urban background concentration as well as global radiation are passed to the OSPM module, which transforms $NO_x$ to $NO_2$ concentrations through chemical processes. For more information about the dispersion and chemical processes of the OSPM see Berkowicz et al. (1997).

*Emission pricing for 'hotspots'*

$NO_2$ concentrations at each receptor point in the study area described in the Scenario section are simulated with the integrated approach described in the previous sections. The resulting concentrations are then compared with the EU limit value for annual mean $NO_2$ concentration of 40 μg/m³ and hourly mean $NO_2$ concentration of 200 μg/m³ (2008/50/EC). Kickhöfer and Nagel (2012) have developed a methodology for the calculation of high-resolution, first-best emission tolls with respect to individual vehicle characteristics and time-dependent link-specific traffic situations. The authors price $CO_2$, NMHC, $NO_x$, PM and $SO_2$ based on average emission cost factors from Maibach et al. (2008)[12] every time a person produces an emission event, or in other words, every time a person leaves a link. For this study, emission costs are assigned to the air pollutants NMHC, $NO_x$, PM and $SO_2$, and the emissions are only priced whenever a person uses one of the links in the network defined as a hotspot. By applying emission costs to 'hotspot links', a direct linkage between polluter and local impacts is drawn. $CO_2$ costs are not considered since the damage costs of $CO_2$ emissions appear globally and cannot be attributed to a certain location. In contrast to $CO_2$, air pollutant concentrations are, to a large extent, directly linked to human health of the residents near the hotspot. Following the methodology by Kickhöfer and Nagel (2012), emission events are converted into monetary terms by an *emission cost module* which uses emission cost factors from Table 5.1 as input. The emission costs for an average passenger car on a local street with a speed limit of 50 km/h are 0.75 EURct/km during free flow and 1.87 EURct/km during stop&go.

**Table 5.1 Emission cost factors by emission type taken from Maibach et al. (2008)**

| Emission type | Cost factor [EUR/ton] |
|---|---|
| **NMHC** | 1,700 |
| **$NO_x$** | 9,600 |
| **PM** | 384,500 |
| **$SO_2$** | 11,000 |

In evaluating the plans, the monetary equivalent of the emission events reduces the utility of a traveller if they drive on one of the links identified in the base case as a hotspot. Additionally, the router module, a time-dependent best path algorithm, uses generalized costs (= disutility of travelling) from the last iteration as input for generating new routes for a certain share of agents. In this chapter, generalized costs on hotspots are in addition to travel time and travel costs dependent on individual emission costs, of which the latter costs depend on vehicle characteristics and time-dependent traffic situations. Therefore, the router also has knowledge about this

12 Emission cost factors in Maibach et al. (2008) are derived from damage cost approaches. Please note that this results in average costs per amount of pollutant; since these amounts are influenced by congestion effects and vehicle attributes, the resulting costs are marginal costs with respect to congestion and vehicle attributes.

highly disaggregated person-specific information. Outside of hotspots, however, a calculation of highly differentiated tolls is not necessary. This improves the performance of the current setup enormously compared to Kickhöfer and Nagel (2012) (by a factor of 10 for the 1 per cent sample of the scenario described in the following section, since it scales with the number of tolled roads).

## Scenario

In this section, we first give a short introduction to the large scale real-world scenario of both the Munich metropolitan and the study area and follow with a description of the simulation approach as well as definition of the available choice dimensions and the utility functions. These sub-sections are followed by a description of the base and policy cases aimed at reducing air pollutant concentration for the hotspots to levels under the EU limit values.

### *Scenario setup*

*Network and population*[13] The road network consists of 17,888 nodes and 41,942 street segments. It covers the federal state of Bavaria, with more detail in and around the city of Munich and less detail further away. Every link is characterized by a maximum speed, a flow capacity and a number of lanes. This information is stored in the road type, which for the emission calculation is always assigned to a corresponding HBEFA road type. In order to obtain realistic time-dependent travel demand data, several data sources have been converted into the MATSim population format. The level of detail from the resulting individual daily plans naturally depends on the information available from either disaggregated stated preference data or aggregated population statistics. Therefore, three 'subpopulations' are created, each corresponding to one of the three different data sources:

- Urban population (based on Follmer et al. (2004)): The synthetic population of Munich is created on the base of very detailed survey data provided by the municipality of Munich RSB (2005), named 'Mobility in Germany' (MiD 2002). Whole activity chains are taken from the survey data for this population. MiD 2002 also provides detailed vehicle information for every household. Linking this data with individuals makes it possible to assign a vehicle to a person's car trip and thus calculate emissions based on this detailed information. However, as of now there is no vehicle assignment module which models intra-household decision making. It is therefore possible that a vehicle is assigned to more than one person at the same time. The synthetic urban population of Munich consists of 1,424,520 individuals.

13 Since the description of network and population generation remains unaltered, this section is taken from Kickhöfer and Nagel (2012).

- Commuter population (based on Böhme and Eigenmüller (2006)): Unfortunately, the detailed data for the municipality of Munich does not contain information about commuters living outside of Munich and working in Munich or about people living in Munich and working outside of Munich. The data analysed by Böhme and Eigenmüller (2006) provides information about workers that are subject to the social insurance contribution with the base year of 2004. With this information, a total of 510,150 synthetic commuters are created from which of 306,160 work in Munich. All commuters perform a daily plan that encodes two trips: from their home location to work and back.
- Freight population (based on ITP/BVU (2005)): Commercial traffic is based on a study published on behalf of the German Ministry of Transport by ITP/BVU (2005). It provides origin-destination commodity flows throughout Germany differentiated by mode and ten groups of commodities. After converting flows that are relevant for the study area into flows of trucks, this population consists of 158,860 agents with one single commercial traffic trip.

Overall, the synthetic population now consists of 2,093,530 agents. To speed up computations, a 1 per cent sample is used in the subsequent simulations. For commuters and freight, no detailed vehicle information is available. Emissions are therefore calculated based on fleet averages for cars and trucks from HBEFA.

*Study area for air pollutant concentration* For the identification of hotspots, an area of approximately 3×3 km² is chosen. This area is located around 'Maxvorstadt', a district north of the Munich city centre with a mixed-use environment of living, working, commercial and cultural activities. Due to such a mixed-use environment, there is a high demand of traffic to and from the area. Figure 5.1 below shows the location of the study area with links of different road types, different street orientation and building structures.

As mentioned above, the air quality module combines information from the transport model, the emission modelling tool, data on street and building geometry, and meteorological data. The data on street and building geometry is provided by the municipality of Munich (RGU 2006). For this chapter, receptor points are placed at the facade of the buildings along street canyons in the study area. Several receptor points are distributed within each street canyon on both sides, resulting in a total of 447 receptor points. Intersections and areas with a large proportion of open space are not included since they show more complex dispersion processes unable to be captured with this approach. From the given data, the following attributes are extracted and used in the OSPM module: from the position of a receptor, the street orientation, street width, building height and all the gaps in the building structure (including the respective height) are calculated. With an extension of ArcGIS, the so-called AirGIS, three shapefiles – street, building and receptor points – are joined and the attributes are calculated (Jensen et al. 2001). An air quality monitoring station, located in the northeast of Munich and operated by the Bavarian

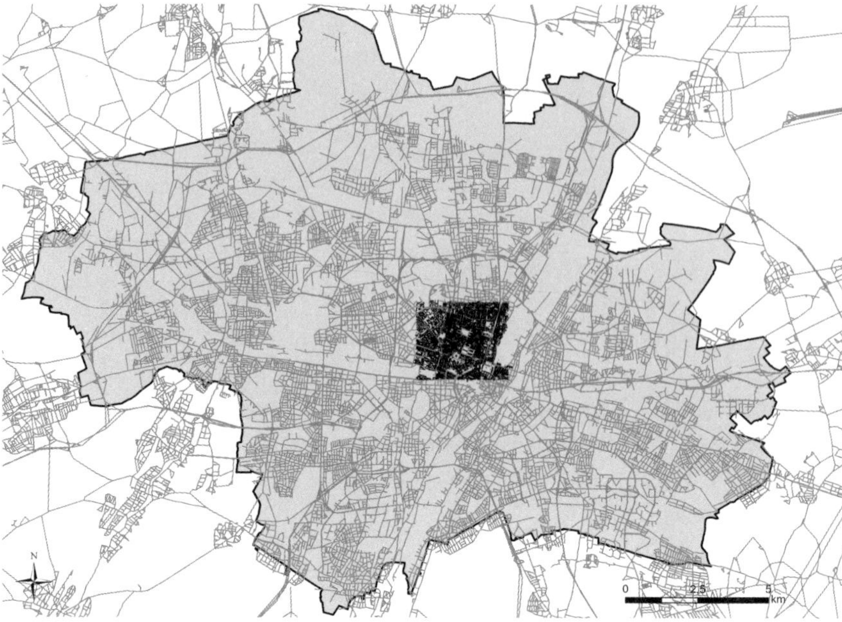


**Figure 5.1 Munich city and study area (black area)**

Environment Agency, measures hourly urban background concentrations of NO, $NO_2$ and $O_3$ (LFU 2008). In addition, hourly meteorological data including wind speed and direction, temperature, and relative humidity are measured at one location in Munich, where the German meteorological service is located (DWD 2008). Data on global radiation are available per month for a grid cell of 1 km (DWD 2008). The calculation of hourly $NO_2$ concentrations is done for the entire year of 2008.

*Simulation*

*Utility-based approach* For the mental layer within MATSim, which describes the behavioural learning of agents, a simple utility-based approach is used in this study. When choosing between different options with respect to a multinomial logit model, agents are allowed to adjust their behaviour among two choice dimensions: route choice and mode choice.[14] The former allows individuals to adapt their routes on the road network when travelling by car. The latter makes it

14 Presumably, there are more choice dimensions in reality such as activity location choice, changes in the frequency of performing activities and changes in bundling activities. A possible approach on how to deal with these possible user reactions within the MATSim framework can be found in Horni et al. (2012). In the section on Results and Discussion, the reader will find a discussion on how the limited choice dimensions in this chapter might influence the results.

possible to change the transport mode for a sub-tour within the agent's daily plan (see Appendix). Only a switch from car to public transit or the other way around is possible. Trips that are initially made with any other mode remain fixed within the learning cycle. From a research perspective, this approach can be seen as defining a system where public transit is a placeholder for all substitutes of the car mode. The following utility functions are used for car and public transit, representing the travel related part of utility[15] (see equation (3) in the Appendix):

$$V_{car,i,j} = -0.96t_{i,car} - 0.07949c_{i,car} \quad (1)$$

$$V_{pt,i,j} = -0.75 - 1.14t_{i,pt} - 0.07949c_{i,pt}, \quad (2)$$

Where $t_i$ is the travel time of a trip to activity $i$ and $c_i$ is the corresponding monetary cost. The following activities are used in this study: work, pick-up, with adult, other, education, business, shopping, private, leisure, sports, home, and friends. Travel times and monetary costs are mode dependent indicated by the indices. The utilities $V_{car,i,j}$ and $V_{pt,i,j}$ for person $j$ are computed in units of utility. Due to a lack of behavioural parameters for the municipality of Munich, these are taken from an Australian study by Tirachini et al. (2012) and adjusted in order to meet the MATSim framework (Kickhöfer and Nagel 2012).

*Simulation procedure* In each of the first 800 iterations, 15 per cent of the agents are forced to discover new routes, 15 per cent change their transport mode for a car or public transit sub-tour in their daily plan and 70 per cent switch between their existing plans. Between iteration 801 and 1,000, route and mode choice are switched off; in consequence, agents only switch between existing options. The output of iteration 1,000 is then used for the identification of hotspots, where all receptor points with a $NO_2$ concentration higher than EU limits[16] are selected.

It is also used as input for the simulation of the following cases:

15 Please note that the following formulas include opportunity costs of time $\beta_{perf}$; the effective Values of Travel Time Savings (VTTS) are therefore 12.08 EUR/h for car and 14.34 EUR/h for public transit. Additionally, the alternative specific constant for public transit was calibrated in order to replicate the modal split in Munich.

16 In this chapter, the emissions are simulated for a representative weekday whereas the EU limit value considers concentrations over an entire year including weekdays and the weekend. In order to reduce this discrepancy, the limit value is modified by the following calculation: based on measured $NO_2$ concentrations over the entire focus year, the relative difference between $NO_2$ concentrations including every day of the week and weekday concentrations were calculated. With the help of this difference, the limit value was then approximated.

- **Base case:** no external cost pricing with unchanged cost structure.
- **Policy cases:** external cost pricing using the methodology from the sub-section on emission pricing for 'hotspots'. In order to capture the impacts of price changes on hotspots, emission cost factors from Table 5.1 are multiplied by constants of 1.0, 10.0, 20.0, 30.0, 40.0, 100.0 and 200.0; this parametric procedure results in seven different policy cases.[17]

User costs[18] for the car are always fixed to 30 EURct/km. For policy cases, additional costs apply (see above). User costs for public transit are assumed to be constant at 18 EURct/km. At this point, all simulations are continued for another 500 iterations. Again, during the first 400 iterations, 15 per cent of the agents perform route adaption while another 15 per cent of the agents choose between car and public transit for one of their sub-tours. The remaining agents switch between existing plans. For the final 100 iterations only a fixed choice set is available for all agents. When evaluating the impact of the two policy measures, the final iteration of 1,500 for every policy case is compared to the iteration of 1,500 for the base case.

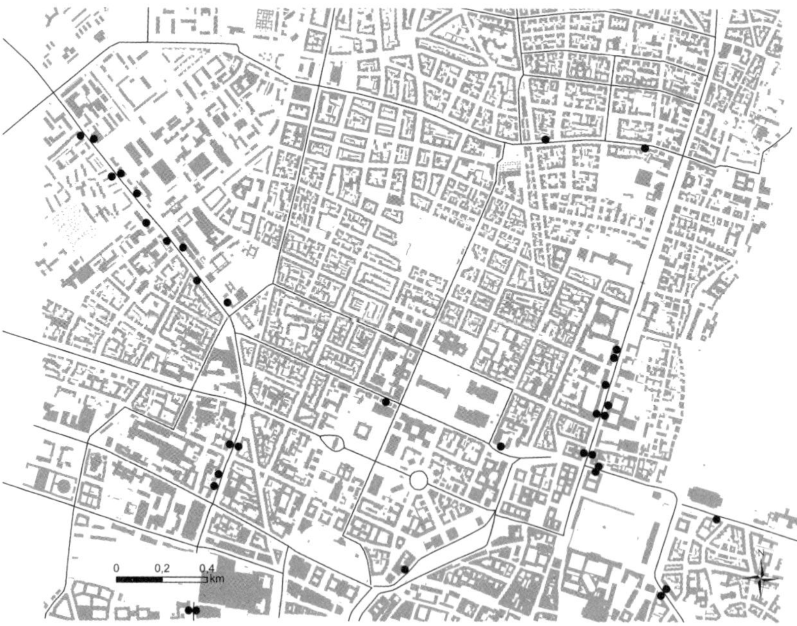


**Figure 5.2 Hotspots in the base case, study area**

17 The highest constant results in emission costs for an average passenger car on a local street with a speed limit of 50km/h of 1.50 EUR/km during free flow and of 3.74 EUR/km during stop&go, according to Maibach et al. (2008).

18 The term 'user costs' refers to the user's out-of-pocket costs.

**Results and discussion**

For the base case, 34 hotspots of the 447 receptor points are identified within the study area. They exceed the annual mean $NO_2$ concentration limit value. These hotspots are mainly located on north-south oriented streets, due to a generally higher traffic volume along those axes and the dominating west wind, leading to a better dilution of air pollution within west-east-oriented street canyons (see Figure 5.2).

Our study found that the hourly mean concentration of $NO_2$ does not exceed the EU limit value of 200 μg/m³ per hour within the study area; this can be explained for several reasons. First, emissions are calculated for one representative weekday averaging out any peak emissions that may occur due to certain circumstances, such as congestion and accidents. The area shows only a few large street segments, not including parts of the middle ring road, where exceedances were measured. Therefore, the following discussion refers only to the annual mean $NO_2$ concentration. Figures 5.3 and 5.4 show the expected effect of different emission cost levels on the number of hotspots and on average annual $NO_2$ concentration. Besides a few exceptions, average annual $NO_2$ concentrations and consequently the number of hotspots decrease with increasing emission costs at the hotspots. Policy case 30[19] shows that the removal of some hotspots can regenerate already eliminated hotspots, which can be mainly explained by an increased choice of routes, including those with no emission costs within the study area, to avoid costly hotspots. Also, the relatively low concentration level reported in policy case 20 can be deduced from a lower share of car trips and longer distances travelled, the latter of which are most likely detours.

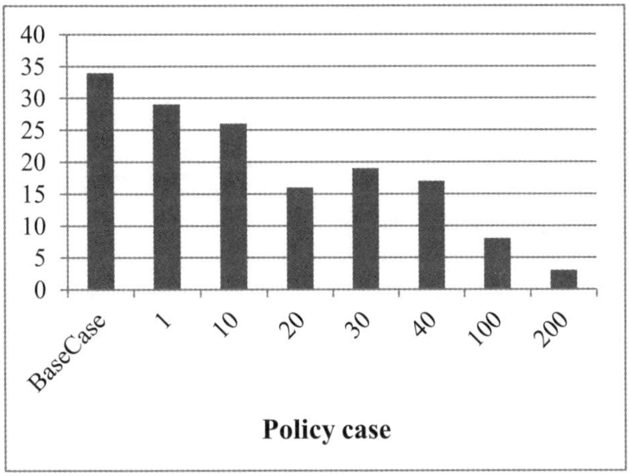


**Figure 5.3 Number of hotspots for each policy scenario**

19 The policy case 30 implies a scenario with emissions costs multiplied by a constant of 30.

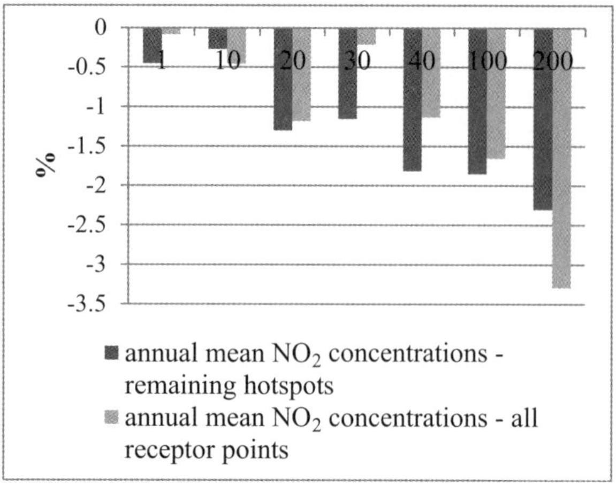


**Figure 5.4 Relative change in annual mean $NO_2$ concentration between policy and base case for the hotspots and for all receptor points**

The reactions of users to policy cases 100 and 200 show a significant decrease in the number of hotspots. Comparing the annual mean concentration of $NO_2$ at all remaining hotspots with the concentration at all receptor points captures the impact of increasing emission costs on the whole study area (see Figure 5.4). The price increase signals a decreasing trend in $NO_2$ concentration at both the remaining hotspots and at all receptor points.

The effect of different pricing schemes on total $NO_2$ emissions in the metropolitan area of Munich is presented in Figure 5.5. Policy cases 20, 30 and 100 show an increase in the emission level as compared with the base case, primarily due to longer car distances travelled. Policy cases 1, 10 and 40 generate a lower emission level compared to the base case. Fluctuations in the relative change of total $NO_2$ emissions, which can be seen as travellers shifting from car to pt and back with increasing costs, presumably result from changes in traffic demand and travel time. Overall though, relative changes in total $NO_2$ emissions are small, indicating a small impact of increasing costs on the entire street network.

The most effective policy case (200) removes almost all hotspots and decreases total $NO_2$ emissions by a small amount while involving a multiple of the emission costs proposed in the literature. This implies an enormous discrepancy between the two approaches of (i) pricing and (ii) the definition of limit values, which can be explained with different reasons. First, this study only sets internalization prices on links with a hotspot. The chosen cost factors from Maibach et al. (2008) are intended to be set on the entire network of links. As a result, travel by car is in general in the simulation more attractive than with a full emission cost internalization strategy (see Kickhöfer and Nagel 2012). Second, this study only allows two user reactions: route choice and mode choice. Therefore, the simulation predicts – for a fixed price change – a lower change in demand than

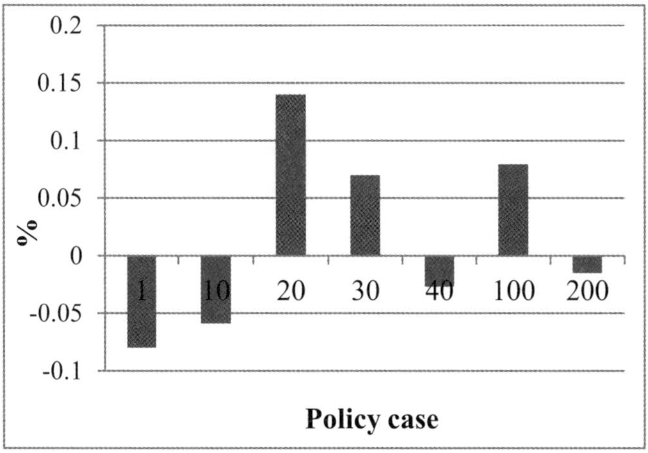


**Figure 5.5 Relative change in total $NO_2$ emissions for each policy case compared to the base case, Munich metropolitan area**

would be observed in reality. In order to remove emission hotspots, this implies higher necessary costs in the simulation.

Third, the calculated high emission costs – 1.50 EUR/km in a free flow traffic situation and 3.74 EUR/km during stop&go – needed to eliminate hotspots may have resulted from possible underestimations of damage costs in the literature. Health impacts of air pollution are especially difficult to determine and some of these effects are likely to be not included in the cost estimation. The exposure response functions are widely derived by epidemiological studies, which reveal several uncertainties, especially for $NO_x$. Long-term studies on exposure to air pollution are time-consuming and costly (Bickel and Friedrich 2005). They often identify higher estimates of health impacts than short-term studies (Künzli et al. 2001). The exposure response functions give an indication of health impacts, but a translation into a willingness to pay for clean air is nontrivial.

The determination of health impacts is not only relevant for an effective pricing strategy though. Whereas Hurley et al. (2005) and Bickel and Friedrich (2005) calculate emission cost factors for $NO_x$ based on indirect health impacts, $NO_2$ limit values are calculated based on health impacts, which are, so far, only related to $NO_2$. Besides the direct effect of $NO_2$, health risks may arise from correlated exhaust components such as ultrafine particles and hydrocarbons, or $NO_2$ chemistry products, including ozone and secondary particles (WHO 2006). According to the WHO (2000: 178), 'there is no evidence for a clearly defined concentration–response relationship for nitrogen dioxide exposure.' The WHO updated the findings on short-term and long-term exposure to $NO_2$ and the identified guidelines from 2000. $NO_2$ levels are regarded as an appropriate marker of exposure to traffic-related emissions. The main challenge in setting up reasonable guidelines is the cocktail of combustion related pollutants $NO_2$ often appears in. If $NO_2$ is used as a marker for this mixture, a lower annual limit value is suggested by the WHO. Other findings also support a lower value, but the existing evidence is not sufficient for revising the limit values (WHO 2006).

In conclusion, $NO_2$ concentrations, which exceed the annual limit value, and potentially even the compliance with the limit value, most likely lead to adverse health effects. The health risks can occur due to exposure to $NO_2$ itself or to a cocktail of pollutants. Therefore, any violation of limit values should be avoided.

Beyond the evaluation of a pricing strategy, the results show that it is crucial to compare pricing measures with limit values on a common framework of assumptions and prerequisites. One step towards this idea, which can be done with this integrated modelling approach, could be, for example, that simulated air pollutant concentrations instead of emissions are used for the approximation of the actual damage costs following the impact pathway approach. This approach considers the calculation of air pollutant emissions, the cumulative exposure from the increased concentrations, the damage from this increase and the monetary valuation, according to Maibach et al. (2008). Such issues should be addressed in future studies. Finally, it is important to note that this study also reveals the enormous challenge in creating a comparable framework for identifying internalization prices and limit values and how they are related.

## Conclusions

This chapter enhances a transport and emission modelling approach by integrating an air quality module. Applying this integrated approach, pricing strategies are evaluated with respect to their mitigation potential in removing $NO_2$ hotspots, which have a higher concentration in Munich than the current EU limit values. Overall, this chapter showed that the effect of increasing emission costs on the number of hotpots shows a decreasing trend. The concentration of $NO_2$ in the study area is reduced and the impact on the entire urban area in terms of total emissions is small. In order to eliminate $NO_2$ hotspots in an urban context, internalizing prices from the literature need to be multiplied by a large number when only being applied on links that exceed the limit values and considering the limitations of this study. Uncertainties remain with respect to the exact cost factor and the limit value though. Having completed the analysis of a singular pricing measure, the following can be concluded: the applied hotspot pricing strategy is effective if user costs are increased considerably, but its efficiency may be questioned. In other words, the internalization price to achieve a social optimum may not go along with the removal of all hotpots. The cause and effect relationship between pricing of external effects and compliance with the environmental limit values is complex. This relation needs to be further investigated, especially the calculation of external costs based on concentrations, direct health impacts and abatement costs, and the comparison with a full emission cost internalization strategy.

In conclusion, a hotspot pricing strategy turns out to be very uncertain in its appropriate formation, especially when it comes to the quantification of the internalization price and its practical implementation. Therefore, additional measures need to be evaluated in future studies. An adequate mixture of measures is necessary with the pricing strategy as one effective policy to sustain the effects on air quality. The current approach allows for the modelling of a package of measure, including, among others, driving bans, speed limits and technological changes. Such an integrated modelling approach can be projected on the entire city of Munich and on other cities given the required data are available.

## Acknowledgements

This work was funded in part by the German Research Foundation (DFG) within the research project 'Detailed evaluation of transport policies using microsimulation'. Important data was provided by the Municipality of Munich, more precisely by 'Kreisverwaltungsreferat München' and 'Referat für Stadtplanung und Bauordnung München'. Our computer cluster is maintained by the Department of Mathematics at Technische Universität Berlin.

## References

Berkowicz, R., Hertel, O., Larsen, S.E., Sorensen, N.N. and Nielsen, M. 1997. *Modelling Traffic Pollution in Streets*. Roskilde: NERI.

Bickel, P. and Friedrich, R. 2005. *Externalities of Energy (ExternE) – Methodology 2005 Update* [Online: Institut für Energiewirtschaft und Rationelle Energieanwendung — IER Universität Stuttgart, Germany]. Available at: http://www.externe.info/externe_d7/sites/default/files/methup05a.pdf [accessed: 28 September 2012].

Böhme, S. and Eigenmüller, L. 2006. *Pendlerbericht Bayern. Technical report*. IAB.

Cetin, N., Burri, A. and Nagel, K. 2003. *A large-scale agent-based traffic microsimulation based on queue model*. Paper to the Swiss Transport Research Conference (STRC): Monte Verita, CH, 2003. Earlier version, with inferior performance values: Transportation Research Board Annual Meeting, Washington, 2003, paper number 03-4272.

Charypar, D. and Nagel, K. 2005. Generating complete all-day activity plans with genetic algorithms. *Transportation*, 32 (4), 369–97.

DWD (Deutscher Wetterdienst) 2008.

EPA 2012. MOVES (Motor Vehicle Emission Simulator) [Online: US Environmental Protection Agency]. Available at: http://www.epa.gov/otaq/models/moves/index.htm [accessed: 27 September 2012].

Follmer, R., Kunert, U., Kloas, J. and Kuhfeld, H. 2004. *Mobilität in Deutschland – Ergebnisbericht. Technical report.* [Online: infas/DIW]. Available at: www.kontiv2002.de [accessed: 27 September 2012].

Gawron, C. 1998. An iterative algorithm to determine the dynamic user equilibrium in a traffic simulation model. *International Journal of Modern Physics C*, 9 (3), 393–407.

Gehlert, T., Kramer, C., Nielsen, O.A. and Schlag, B. 2011. Socioeconomic differences in public acceptability and car use adaptation towards urban road pricing, *Transport Policy*, 18, 685–94.

Gerike, R., Gehlert, T., Richter, F. and Schmidt, W. 2008. Think global, act local – the internalisation of external costs in transport. *European Transport \ Trasporti Europei*, 38, 61–84.

Grether, D., Chen, Y., Rieser, M. and Nagel, K. 2009. Effects of a simple mode choice model in a large-scale agent-based transport simulation, in *Complexity and Spatial Networks. In Search of Simplicity*, edited by A. Reggiani and P. Nijkamp. Berlin: Springer, 167–86.

Hatzopoulou, M. and Miller, E.J. 2010. Linking an activity-based travel demand model with traffic emission and dispersion models: transport's contribution to air pollution in Toronto. *Transportation Research Part D: Transport and Environment*, 15 (6), 315–25.

Holmes, N.S. and Morawska, L. 2006. A review of dispersion modelling and its application to the dispersion of particles: an overview of different dispersion models available. *Atmospheric Environment*, 40, 5902–28.

Horni, A., Nagel, K. and Axhausen, K. 2012. *High-resolution destination choice in agent-based demand models*. Annual Meeting Preprint 12-1989: Transportation Research Board, Washington D.C.

Hülsmann, F., Gerike, R., Kickhöfer, B., Nagel, K. and Luz, R. 2011. Towards a multi-agent based modeling approach for air pollutants in urban regions, in *Luftqualität an Straßen*, Bergisch Gladbach: Bundesanstalt für Straßenwesen, FGSV Verlag GmbH, 144–166.

Hurley, F., Hunt, A., Cowie, H., Holland, M., Miller, B., Pye, S. and Watkiss, P. 2005. *Methodology for the Cost-Benefit analysis for CAFE: Volume 2: Health Impact Assessment, Service Contract for carrying out cost-benefit analysis of air quality related issues, in particular in the clean air for Europe (CAFE) programme*. Didcot: AEA Technology Environment.

INFRAS 2010. *Handbook Emission Factors for Road Transport 3.1. Technical report* [Online]. Available at: www.hbefa.net [accessed: 27 September 2012].

ITP/BVU 2005. *Prognose der deutschlandweiten Verkehrsverflechtungen 2025. Technical report* [Online]. Available at: http://daten.clearingstelle-verkehr.de/220/ [accessed: 27 September 2012].

Jensen, S.S., Berkowicz, R., Hansen, H.S. and Hertel, O. 2001. A Danish decision-support GIS tool for management of urban air quality and human exposures. *Transportation Research Part D: Transport and Environment*, 6 (4), 229–41.

Kakosimos, K.E., Hertel, O., Ketzel, M. and Berkowicz, R. 2010. Operational Street Pollution Model (OSPM) – a review of performed validation studies and future prospects. *Environmental Chemistry*, 7, 485–503.

Ketzel, M., Berkowicz, R., Hvidberg, M., Jensen, S.S. and Raaschou-Nielsen, O. 2012. Evaluation of AirGIS – a GIS-based air pollution and human exposure modelling system. *Intl. J. Environ. Poll.* In press.

Kickhöfer, B., Grether, D. and Nagel, K. 2011. Income-contingent user preferences in policy evaluation: application and discussion based on multi-agent transport simulations. *Transportation*, 38, 849–70.

Kickhöfer, B., Hülsmann, F., Gerike, R. and Nagel, K. 2012 (in press). Rising car user costs: comparing aggregated and geo-spatial impacts on travel demand and air pollutant emissions, in *Smart Transport Networks: Decision Making, Sustainability and Market Structure. NECTAR Series*, edited by T. Vanoutrive and A. Verhetsel. Cheltenham: Edward Elgar.

Kickhöfer, B. and Nagel, K. 2012. *Using high-resolution first-best tolls as a benchmark for policy evaluation: the case of air pollution costs.* Paper to the Kuhmo Nectar Conference on Transportation Economics: Berlin, June 2012.

Künzli, N., Medina, S., Kaiser, R., Quenel, P., Horak, F. Jr. and Studnicka, M. 2001. Assessment of deaths attributable to air pollution: should we use risk estimates based on time series or on cohort studies? *Am J Epidemiol*, 153, 1050–55.

Lefebvre, N. and Balmer, M. 2007. *Fast shortest path computation in time-dependent traffic networks*. Paper to the Swiss Transport Research Conference (STRC): Monte Verita, CH, September 2007.

LFU (Bavarian Environment Agency) 2008.

LFU (Bavarian Environment Agency) 2012. Überschreitungshäufigkeit für Feinstaub-PM10, Stickstoffdioxid und Schwefeldioxid. *Technical report* [Online: Landesamt für Umwelt]. Available at: http://www.lfu.bayern.de/luft/lueb/ueberschreitung_pm10_so2_no2/index.htm [accessed: 27 September 2012].

Maibach, M., Schreyer, D., Sutter, D., van Essen, H., Boon, B., Smokers, R., Schroten, A., Doll, C., Pawlowska, B. and Bak, M. 2008. *Handbook on estimation of external costs in the transport sector. Technical report*. Delft: CE Delft.

Raney, B. and Nagel, K. 2006. An improved framework for large-scale multi-agent simulations of travel behaviour, in *Towards Better Performing European Transportation Systems*, edited by P. Rietveld, et al. London: Routledge, 305–47.

RGU (Municipality of Munich: Referat für Gesundheit und Umwelt) 2006.

Rieser, M., Grether, D. and Nagel, K. 2009. Adding mode choice to a multi-agent transport simulation. *Transportation Research Record: Travel Demand Forecasting 2009*, 2132, 50–58.

RSB (Municipality of Munich: Referat für Stadtplanung und Bauordnung) 2005.

Samaras, Z. and Zierock, K.-H. 2009. *Methodology for the calculation of exhaust emissions – SNAPs 070100-070500, NFRs 1A3bi-iv* [Online: EEA]. Available at: http://www.emisia.com/documentation.html [accessed: 27 September 2012].

Smit, R., Smokers, R. and Rabe, E. 2007. A New Modelling Approach for Road Traffic Emissions: VERSIT + LD – Background and Methodology. *Transportation Research Part D: Transport and Environment*, 12 (6), 414–22.

Tirachini, A., Hensher, D.A. and Rose, J.M. 2012. *Multimodal pricing and optimal design of public transport services: the interplay between traffic congestion and bus crowding*. Paper to the Kuhmo Nectar Conference on Transportation Economics: Berlin, June 2012.

Vardoulakis, S., Fisher, B.E.A, Pericleous, K. and Gonzales-Flesca, N. 2003. Modelling air quality in street canyons: a review. *Atmospheric Environment*, 37, 155–82.

WHO Regional Office for Europe 2000. *Air quality guidelines for Europe, 2nd ed.* [Online: World Health Organisation]. Available at: http://www.euro.who.int/en/what-we-publish/abstracts/air-quality-guidelines-for-europe [accessed: 23 October 2012].

WHO Regional Office for Europe 2006. *Air Quality Guidelines Global. Update 2005. Particulate matter, ozone, nitrogen dioxide and sulfur dioxide* [Online: World Health Organisation]. Available at: http://www.euro.who.int/en/what-we-publish/abstracts/air-quality-guidelines.-global-update-2005.-particulate-matter,-ozone,-nitrogen-dioxide-and-sulfur-dioxide [accessed: 23 October 2012].

Zallinger, M.S., Luz, R., Hausberger, S., Hirschmann, K. and Fellendorf, M. 2010. *Coupling of microscale traffic and emission models to minimize emissions by traffic control systems*. Paper to the 18th International Symposium: Transport and Air Pollution, Dübendorf, 2010, 237–43.

## Appendix[20]

The following paragraphs are meant to present more information about the MATSim simulation approach that is used in this chapter. Every step of the iterative loop in the MATSim sub-section is now illustrated in more detail.

*Plans generation*

An agent's daily plan contains information about his planned activity types and locations, about duration and other time constraints of every activity, as well as the mode, route, the desired departure time and the expected travel time of every intervening trip (= leg). Initial plans are usually generated based on microcensus information and/or other surveys. The plan that was reported by an individual is in the first step marked as 'selected'.

20 Since the methodology remains unaltered, this section is taken from Kickhöfer and Nagel (2012).

*Traffic flow simulation*

The traffic flow simulation executes all selected plans simultaneously in the physical environment and provides output describing what happened to each individual agent during the execution of its plan. The car traffic flow simulation is implemented as a queue simulation, where each road (= link) is represented as a first-in first-out queue with two restrictions (Gawron 1998, Cetin et al. 2003): first, each agent has to remain for a certain time on the link, corresponding to the free speed travel time. Second, a link storage capacity is defined which limits the number of agents on the link; if it is filled up, no more agents can enter this link. The public transport simulation simply teleports agents between two activity locations. The distance is defined by a factor of 1.3 times the beeline distance between the locations. Travel speed can be configured and is set in this chapter to 25 km/h. Public transit is assumed to run continuously and without capacity restrictions (Grether et al. 2009, Rieser et al. 2009). All other modes are modelled similar to public transport: travel times are calculated based on mode specific travel speed and the distance estimated for public transport. However, the attributes of these modes are not relevant for this chapter since agents are only allowed to switch from car to public transport and the other way around. Trips from the survey that are not car or public transport trips, remain fixed during the learning cycle. Output of the traffic flow simulation is a list that describes for every agent different events, for example entering or leaving a link, arriving or leaving an activity. These events are written in XML-format and include agent ID, time and location (link or node ID). It is, therefore, quite straightforward to use this disaggregated information for the calculation of link travel times or costs (which is used by the router module), trip travel times, trip lengths, and many more.

*Evaluating plans*

In order to compare plans, it is necessary to assign a quantitative measure to the performance of each plan. In this work, a simple utility-based approach is used. The elements of our approach are as follows:

- The total utility of a plan is computed as the sum of individual contributions:

  $$V_p = \sum_{i=1}^{n} (V_{perf,i} + V_{tr,i}), \tag{3}$$

  where $V_p$ is the total utility for a given plan; $n$ is the number of activities; $V_{perf,i}$ is the (positive) utility earned for performing activity $i$; and $V_{tr,i}$ is the (usually negative) utility earned for travelling during trip $i$. Activities are assumed to wrap around the 24-hours-period, that is, the first and the last activity are stitched together. In consequence, there are as many trips between activities as there are activities.

- A logarithmic form is used for the positive utility earned by performing an activity (see, for example, Charypar and Nagel 2005, Kickhöfer et al. 2011):

$$V_{perf,i}(t_{perf,i}) = \beta_{perf} t_{*,i} \ln(\frac{t_{perf,i}}{t_{0,i}}) \quad (4)$$

  where $t_{perf}$ is the actual performed duration of the activity, $t_{*,i}$ is the 'typical' duration of activity $i$, and $\beta_{perf}$ is the marginal utility of an activity at its typical duration. $\beta_{perf}$ is the same for all activities, since in equilibrium all activities at their typical duration need to have the same marginal utility. $t_{0,i}$ is a scaling parameter that is related both to the minimum duration and to the importance of an activity. As long as dropping activities from the plan is not allowed, $t_{0,i}$ has essentially no effect.
- The disutility of travelling used for simulations is taken from Tirachini et al. (2012). More details are given in the Simulation sub-section.

In principle, arriving early or late could also be punished. For this chapter however, there is no need to do so, since agents are not allowed to reschedule their day by changing departure times. Arriving early is already implicitly punished by foregoing the reward that could be accumulated by doing an activity instead (opportunity cost). In consequence, the effective (dis)utility of waiting is already $-\beta_{perf} t_{*,i} / t_{perf,i} \approx \beta_{perf}$. Similarly, this opportunity cost has to be added to the time spent travelling.

*Learning*

After evaluating daily plans in every iteration, a certain number of randomly chosen agents are forced to re-plan their day for the next iteration. This learning process is, in this study, made by two modules corresponding to the two choice dimension available: a module called *router* for choosing new routes on the road network and a module called *sub-tour mode choice* for choosing a new transport mode for a car or public transport trip. The router module bases its decision for new routes on the output of the car traffic flow simulation and the knowledge of congestion in the network. In the policy cases, it also uses the knowledge about expected emission costs on hotspots. The router is implemented as a time-dependent best path algorithm Lefebvre and Balmer (2007), using generalized costs (= disutility of travelling) as input. The sub-tour mode choice module changes the transport mode of a car sub-tour to public transport or from a public transport sub-tour to car. A sub-tour is basically a sequence of trips between activity locations. However, the simulation needs to make sure that a car can only be used if it is parked at the current activity location. Thus, a sub-tour is defined as a sequence of trips where the transport mode can be changed while still being consistent with the rest of the trips. It is for example assured that a car which is used to go from home to work in the morning needs to be back at the home location in the evening.

If the car remains, for example, at the work location in order to use it to go for lunch, then the whole sub-tour of going to work and back needs to be changed to public transport.

# Chapter 6
# Do Parking Measures Contribute to a Sustainable Transportation System?

Peter van der Waerden and Harry Timmermans

## Introduction

One of the biggest challenges urban transportation planners have is to optimize the accessibility of urban areas and activities in order to better the quality of life in and around these areas. One group of measures planners have at their disposal concerns parking. With the use of parking measures, planners can regulate the use of parking facilities or reduce car use in favour of more environmentally friendly travel modes such as public transport and the bicycle. Litman (2006: 4) describes the benefits of parking management for mobility management, seeing parking management as '*an important component of efforts to encourage more efficient transportation patterns*'. The effects of parking measures on traffic flows and the related topic of emissions are still unclear though. For example, Marsden (2006) provided an overview of parking studies investigating the effects of parking measures in relation to the following different policy contexts: commuter, non-commuter and residential parking. After summarizing the studies, he concluded that it is still difficult to get a clear view of the effects of parking measures. This holds especially for non-commuters (for example, shoppers) who have more freedom in changing their travel behaviour. Based on research identifying the impact of parking measures on energy consumption and air pollution, Höglund (2004) concluded that environmental issues related to car parking should be given much more attention in town and city planning.

Examples of parking measures that can be considered by planners include increasing the cost of parking, relocating parking facilities and closing parking facilities. A major question when implementing one or more parking measures is 'What are the effects of the measure on the various choices travellers make when they go shopping?' Such measures can stimulate travellers to change travel mode, destination, parking location and even departure time and travel company. In addition, one can ask 'Does a change in one or more travel choices affect car use and/or distance travelled by car?' The latter question is the focus of the study described in this chapter. The aim of this chapter is to show some of the effects of parking measures on shoppers' car use and distance travelled by car in the context of shopping trips. Its application is based on a hypothetical situation.

The remainder of the chapter is organized as follows. First, different parking measures and their effects on travel behaviour and, in addition, on energy consumption and air pollution are described. Next, the adopted research approach is outlined. This part is followed by the presentation of the research findings. The chapter ends with some conclusions.

## Effects of parking measures

A variety of studies concerning the effects of parking measures are available (for example, Hensher and King 2001, Farrel and Mahony 2005, Shiftan and Golani 2005, Beunen et al. 2006, Van der Waerden 2012, Danwen et al. 2010). In general, these studies can be divided into two groups: studies that relate parking to environmental issues and studies that relate parking to travel behaviour. Additionally, the first group of studies can be subdivided into studies that focus on the relationship between environmental issues and parking facility design (for example, Davis et al. 2010, Gallagher et al. 2011) and studies that focus on the relationship between environmental issues and traffic flows generated by parking facilities (for example, Höglund 2004, CROW 2010).

Parking facility design can be related to several environmental issues including land use and pollution. To illustrate this relationship, two recent studies are described in more detail. In the first study, Davis et al. (2010) investigated the space allocation of parking lots in relation to the loss of ecosystem services (represented by the area of the parking lots). To value the loss of ecosystem services due to the presence of parking lots in an area, they estimated an ecosystem service value (ESV). In the county considered by Davis et al., the ESV is equal to \$58.6 million. Therefore, if the entire county's parking lots were replaced by wetlands, the ESV of the county would be \$22.5 million – an increase of approximately 38 per cent of the county's ESV. In addition, Davis et al. assessed the amount of runoff and contaminant loads produced by parking lots using the Long-Term Hydrologic Impact Assessment (L-THIA) model. The outcomes of this model show a high amount of runoff and NPS pollutants in relation to parking lot developments.

Gallagher et al. (2011) investigated the effect of three configurations of parking systems (parallel, perpendicular and 45° parking) on pedestrian pollutant exposure. The study was based on the idea that the parking system itself could be used as a passive control of air pollution exposure in street canyons. The results showed that parallel parked cars provided the best overall results with a minimum reduction of 31 per cent and a reduction of up to 49 per cent in varying wind conditions. Here, the parked cars act as a kind of 'baffle-plate' between traffic emissions and pedestrians on the footpath. The fraction of cars parked also influenced the level of reduction of pollutants on the footpath.

In 2010, CROW presented a study regarding the relationship between parking and air quality, giving some insights into the effect of parking measures on the air quality in central business areas. Three different categories concerning the effects of parking measures are investigated: effects on size and composition of the urban fleet of cars, effects on the urban traffic volume (travel kilometres) and effects on emissions per (car) kilometre. Different parking measures are evaluated for their reduction of $NO_2$ (nitrogen dioxide) and $PM_{10}$ (particulate matter). The calculated effects are based on 'parking elasticity' extracted from international literature. Looking at individual parking measures, the biggest reduction of emissions in central business areas can be achieved by a creating a parking ban for heavily polluting cars and by a reducing parking spaces by 25 per cent. As shown, combining parking measures results in higher reductions.

Höglund (2004) examined the impacts of different ways of parking on emissions and air pollution. Based on two traffic models (regional model BRAHE and local model SimPark), the impacts of two ways of parking in Stockholm are compared: curbside parking and underground parking garage. BRAHE calculates regional environmental effects based on changes in local parking conditions (location and type of parking facilities). The effects are expressed in travelled distances. Höglund found that parking in an underground garage results in less emission than curbside parking.

In addition to the studies mentioned above, several studies have dealt with relationships between parking policies, including both parking measures and individual travel choice. For example, Litman (2006) presented a complete overview of different parking policies and their typical effects on required parking spaces and traffic volumes. Table 6.1 shows a limited set of possible parking policies, the percentage reduction in required spaces they typically provide and whether they also tend to reduce traffic problems. The table indicates parking strategies that typically reduce required parking spaces and whether each strategy helps to reduce vehicle traffic and therefore provide additional benefits. Litman (2006: 25) warns that '*not every strategy is appropriate in every situation*'. The success of a parking measure strongly depends on geographic and demographic factors, how a strategy is implemented, and other factors such as available travel mode alternatives, financial incentives and complementary initiatives.

The Dutch organization CROW (2001) investigated the effects of introducing paid parking in major shopping areas in several cities. Based on 18 cases, five regional parking studies and an expert meeting, the institute formulated the following conclusion regarding paid parking. The investigated studies were carried out by different organizations such as municipalities, consultancies and universities. They found that the introduction of paid parking leads to an improvement in the parking climate in town and city centres. The introduction of the policy usually increased parking pressure in areas where parking charges had not yet been introduced. Fees were largely associated with positive effects on the local economy over the long term, although over the short term there might have been a drop in the number of visitors to such areas. This shift from negative to positive effects did not only occur over a matter of years (long-term acceptance), but also occurred as a result of extra measures to increase the attractiveness of the shopping area, such as new shops and the renovation of existing shopping areas. In relation to the parking process, parking fees produced some additional benefits: shorter time periods when looking for parking spaces, more efficient use of parking spaces and the promotion of 'short stay' parking. The CROW (2001) concluded that customers are more likely to change their shopping destination than to change travel mode.

**Table 6.1 Partial overview of different parking policies, derived from Litman (2006)**

| Policy | Description | Parking demand reduction | Traffic reduction |
|---|---|---|---|
| Parking regulations | Apply regulations that favour higher-value use such as service vehicles, deliveries, customers and people with disabilities | 10–30% | |
| Bicycle facilities | Provide bicycle storage and changing facilities | 5–15% | ✓ |
| Parking pricing | Charge motorists directly for using parking facilities | 10–30% | ✓ |
| Improve user information | Provide convenient information on parking availability and price using maps, signs, brochures and the Internet | 5–15% | ✓ |
| Improve enforcement | Insure that enforcement is efficient, considerate and fair | Varies | |

## Research approach

To assess the effects of parking measures on travellers' destination, mode and parking choice behaviour, the Urban Planning Group at Eindhoven University of Technology developed a parking analysis tool called ***Pamela*** (which stands for a ***P****arking* ***A****nalysis* ***M****odel for predicting* ***E****ffects of* ***L****ocal* ***A****reas*) (Van der Waerden 2012). The tool integrates various discrete choice models in order to describe the composition of travellers' parking consideration sets, travellers' combined travel choices of destination, travel mode, parking/bike stall selection and car drivers' adaptive parking choice when faced with a fully occupied parking facility. Several mixed logit models are estimated. The estimated models include various characteristics of shopping destinations (supply and distribution of shops), travel modes (travel time of car, bike and bus), parking (such as the number of parking spaces, parking costs, walking distance and maximum parking duration) and bicycle stall facilities (including security level, storage charge and walking distance) which can be used to define various planning scenarios. Further details of the data collection, model specification and estimation process can be found in Van der Waerden (2012).

To evaluate different transport policies, all models of ***Pamela*** are implemented in the multi-agent platform of NetLogo (Wilensky 1999). NetLogo is an open source, programmable and multi-agent modelling platform. The platform is not designed for specific problems, but rather allows for the definition of almost any problem of interest through programming. NetLogo also provides a graphical representation referred to as 'world', which is especially useful for visualizing spatial activities. World is a grid-based graph composed of cells (patches) with a definable size where an agent (turtle) stand and moves. The agent movement simulation is controlled by commands that adjust the status of 'patches' and the activities of 'turtles'. NetLogo is designed to be object-oriented, which means that patches and turtles may contain properties implemented as variables and behaviours, seen as programmed procedures. It simulates the real world by representing changes in the status of multiple agents 'simultaneously' (of course, processed sequentially in computation) as if they occur in parallel. Interactions between agents in the sense of passing information to one another can also be modelled. This platform is used here to evaluate different parking measures in the context of shopping.

The application of this study consists of a hypothetical situation in the context of shopping (Figure 6.1) and is detailed as follows. The roads and building blocks follow a regular pattern of cells. Roads have a width of 1 cell and building blocks have a width and a length of 4 cells. In line with the content of the combined travel choice model of ***Pamela***, shopping centres (building symbol) are defined by the two characteristics: supply and distribution of shops. Parking facilities (car symbol) are defined using characteristics included in ***Pamela***: size, chance of finding free space, parking costs, maximum parking duration, average egress time, driving space, type of facility, type of security, location in relation to residents' home location, location in relation to other parking facilities and distance to the nearest supermarket/department store. Bicycle stalls (bicycle symbol) are described using different characteristics: level of security, storage charge and walking distance to nearest supermarket or department store. As shown in Figure 6.1, three shopping centres are defined. In addition, nine parking facilities are included. Parking facilities 1, 2 and 3 are located around shopping centre 1, while parking facilities 4 and 5 are located close to shopping centre 2. Parking facilities 6 through 9 are close to shopping centre 3. The bike stall numbers correspond to the number of each shopping centre. Additionally, the situation does not have a scale and it only includes the relative locations of the residents' homes and other various facilities (shopping centre, parking facilities and bicycle stalls).

Shopping trips consisting of 500 residents are simulated during a single day period using the various models of ***Pamela***. The simulation includes the whole process from the generation of a traveller who goes shopping (Step 1) until the traveller returns from the shopping centre to their home location (Step 8, Figure 6.2).

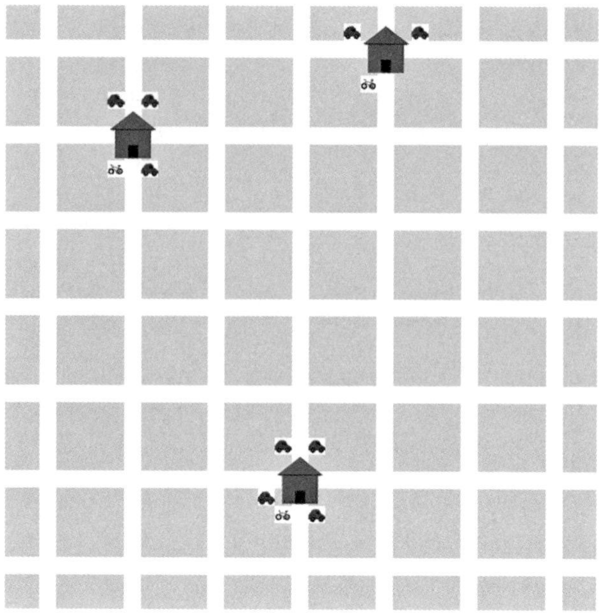

**Figure 6.1 Hypothetical environment of the simulation**

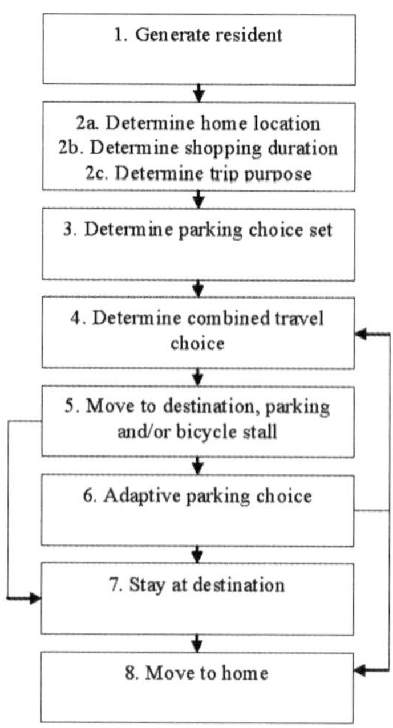


**Figure 6.2 Flow chart of the simulation**

To cancel out random effects, a total of 10 simulation runs were carried out started with a different random seed. The residents wanting to shop are defined using the following characteristics: home location, distance between home location and shopping centres, type of shopper (dark for weekly shopper and light for non-weekly shoppers), shopping duration and location of various parking facilities vis-à-vis the shoppers' home location (Figure 6.3). The distance (D) between home location *i* (coordinates: $x_i$ and $y_i$) and the available shopping centre *j* (coordinates: $x_j$ and $y_j$) are determined for each resident using a rectangular road network and the following formula:

$$D_{ij} = |x_i - x_j| + |y_i - y_j|$$

The various choices a traveller has to make are based on the traveller's characteristics and the characteristics of the destinations, travel modes, parking facilities/bike stalls, and the choices previous travellers make. This final influence is important as the occupancy of parking facilities and the number of cars waiting for a free spot affects traveller choice, all of which are included in the adaptive parking choice model.

Two kinds of transport policies are evaluated. The first policy (labelled **OneEuro**) consists of standardizing parking costs at all parking facilities at 1.00 euro per hour. The policy includes increasing parking costs at two parking facilities from free to 1.00 euro as well as increasing costs at four parking facilities from 0.50 to 1.00 euro. The second policy (labelled **ParkDistance**) focuses on the walking distance between parking and the nearest supermarkets (weekly shopping) or department stores (non-weekly shopping). With different infrastructure measures, such as breakthroughs and movement of parking spaces, every walking distance is set to 150 metres. Based on this hypothetical case, the effects of both transport policies are calculated in relation to a certain base situation (labelled **Base**). This simulation provides an insight into any changes in the composition of parking consideration sets, destination, travel mode, parking or bike stall choices, adaptive parking choices and travel distances. In relation to sustainability, special attention is paid to changes in residents' car use in combination with their destination choice. In addition, travel distance by car is also considered as an important cause of emission rates.

The outcome of the simulation is presented for the entire day (between 8:00 and 20:00 hours) as considered in the simulation. The figures present the results of the residents' destination and travel mode choices.

The effects of the suggested transport policies on shoppers' travel mode choice are presented in Figure 6.4. According to Figure 6.4, it appears that the effect of the investigated transport policies on travel mode choice is considerable (with a maximal change of almost 30 per cent). As expected, the equalization of parking costs produces a shift in travel mode choice from the car (with a decrease of almost 30 per cent of all residents) to the bicycle (with an increase of approximately 27 per cent) and the bus (with an increase of 3 per cent). It also seems that the

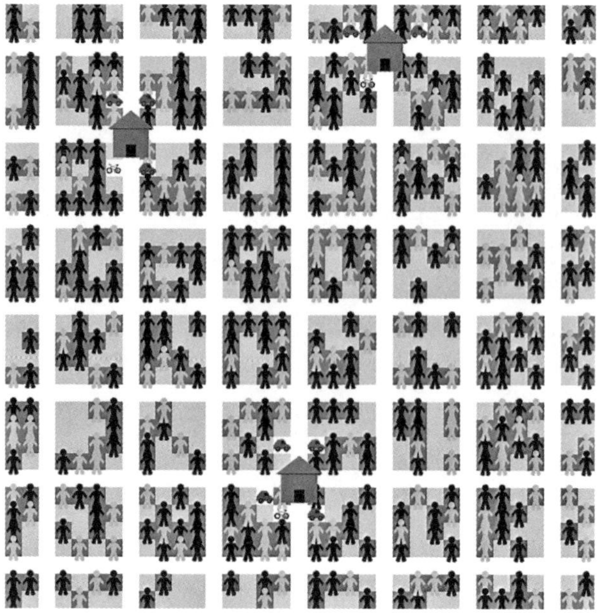

**Figure 6.3 Simulation of 500 residents during day time period results**

relocation of parking facilities marginally affects resident mode choice. Car use decreases by almost 4 per cent while bike use increases with the same percentage. Bus use remains more or less the same in both situations.

The other outcome of the simulation concerns the distance travelled by travel mode. Figure 6.5 presents the total distance travelled by 500 residents (averaged over 10 simulation runs). The formula presented in the previous section has been used to calculate the travel distance between the resident's home location (see dark grey squares in Figure 6.3) and the chosen destination (shopping centre and, in addition, parking facility, bike stall or bus stop). Figure 6.5 shows that the total distance travelled by car in the case of the equalization of parking costs (OneEuro) decreases with more than 35 per cent (from 9200 to approximately 5900 units). In addition, the distance travelled by bike increases with almost 150 per cent (from approximately 1700 to 4200 units). In line with previous findings, the relevance of the bus is marginal. The relocation of parking facilities (ParkDistance) results in a decrease in the total distance travelled by car to approximately 500 units and an increase in the total distance travelled by bike to almost 500 units.

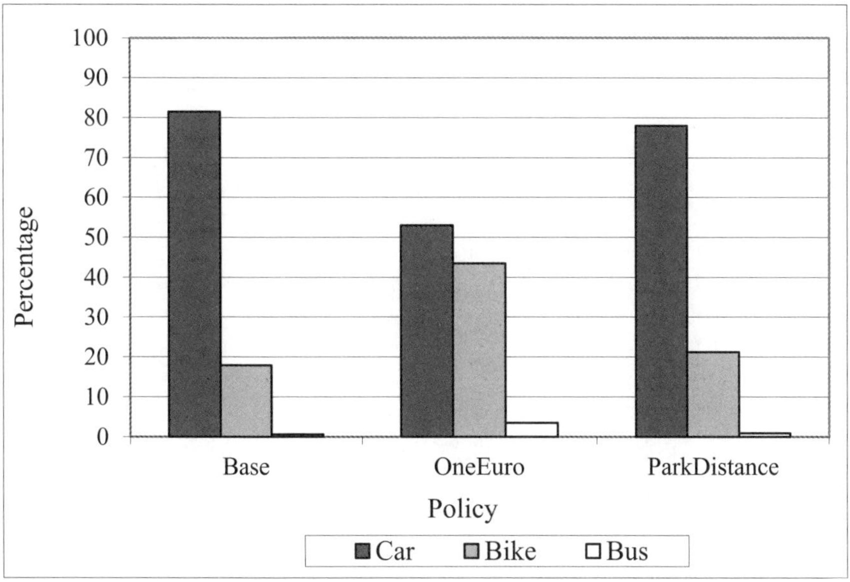


**Figure 6.4 Effects of transport policies on travel mode choice**

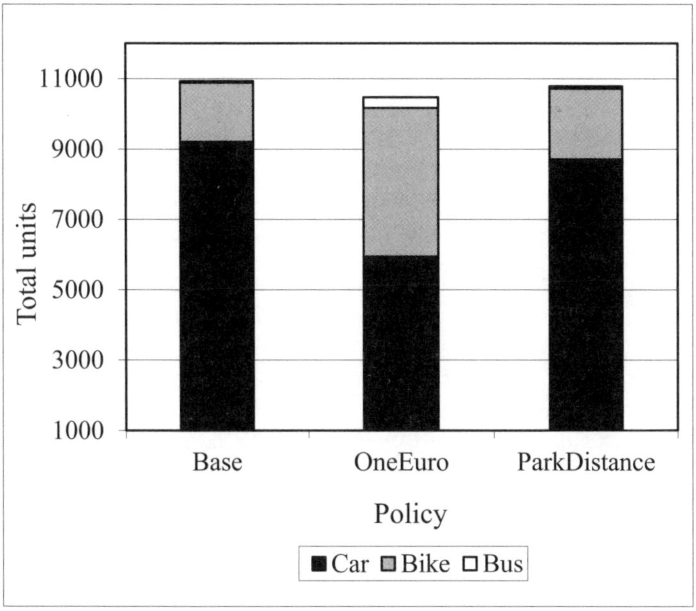


**Figure 6.5 Effects of transport policies on total distance travelled**

## Conclusions

This chapter presented an illustration of the way parking measures might affect car use and distance travelled by cars. The illustration was set up for residents' shopping trips. The parking analysis model ***Pamela*** was used to estimate the effects of two different transport policies: the equalization of parking costs at the level of 1.00 euro per hour and the relocation of parking facilities at a walking distance of 150 metres between parking and final destination. The first policy resulted in a considerable decrease in car use in favour of bicycle use, while the second policy only had a marginal effect on the travel mode choice of shoppers. The same holds for the total distance travelled between home and shopping location and vice versa. Despite the fact that this study is set up as an illustration, the results show that, as expected, price measures are more influential than 'infrastructural' measures and that parking measures affect car use in terms of number of cars and distance travelled by car.

One of the most interesting extensions of the approach presented in this chapter would consist of applying the approach to a real world situation (in practice) and including detailed information of residents' home location, their shopping behaviour as well as characteristics of shopping centres, parking facilities and bicycle stalls. This simulation could be set up to investigate the competition between shopping centres at different spatial levels such as the neighbourhood level (different neighbourhood shopping centres), the city level (main versus different sub-centres) and the regional level (different city centres). Also, linking such research with fuel consumption and pollution production models would be necessary when introducing the approach in practice.

## References

Beunen, R., Jaarsma, C.F. and Regnerus, H.D. 2006. Evaluating the effects of parking policy measures in nature areas. *Journal of Transport Geography*, 14, 376–83.

CROW 2001. *Effects of Parking Measures*, COST 342, CORDIS.

CROW 2010. *Parking and Air Quality, Parking Measures in Central Business Areas and P+R facilities* (in Dutch). Ede: CROW, Nationale Kennisplatform voor Infrastructuur, Verkeer, Vervoer en Openbare ruimte.

Danwen, B., Wei, D. and Shishui, G. 2010. Impact of parking rates on resident travel behaviour. *Journal of Transportation Systems Engineering and Information Technology*, 10, 80–85.

Davis, A.Y., Pijanowski, B.C., Robinson, K. and Engel, B. 2010. The environmental and economic costs of sprawling parking lots in the United States. *Land Use Policy*, 27, 255–61.

Farrel, S. and Mahony, M. 2005. *Attitudes and Behavioral Responses to Measures to Deal with Workplace Parking: A Case Study of Dublin, Ireland.* Compendium of Papers CD-ROM of the 82nd Annual Meeting of the Transportation Research Board, Washington DC, 2005.

Gallagher, J., Gill, L.W. and McNabola, A. 2011. Optimizing the use of on-street car parking system as a passive control of air pollution exposure in street canyons by large eddy simulation. *Atmospheric Environment*, 45, 1684–94.

Hensher, D.A. and King, J. 2001. Parking demand and responsiveness to supply, pricing and location in the Sydney central business district. *Transportation Research Part A*, 35, 177–96.

Höglund, P.G. 2004. Parking, energy consumption, and air pollution. *Science of the Total Environment*, 334–5, 39–45.

Litman, T. 2006. *Parking Management, Strategies, Evaluation, and Planning* [Online: Victorian Transport Policy Institute]. Available at: www.vtpi.org/park_man.pdf [accessed: 20 May 2012].

Marsden, G. 2006. The evidence base for parking policies: a review. *Transport Policy*, 13, 447–57.

Shiftan, Y. and Golani, A. 2005. *The Effect of Auto Restrain Policies on Travel Behavior.* Compendium of Papers CD-ROM of the 82nd Annual Meeting of the Transportation Research Board, Washington DC, 2005.

Van der Waerden, P. 2012. ***Pamela,*** *a* ***P****arking* ***A****nalysis* ***M****odel for predicting* ***E****ffects of* ***L****ocal* ***A****reas*. Eindhoven: Eindhoven University of Technology.

Wilensky, U. 1999. *NetLogo*. Available at: http://ccl.northwestern.edu/netlogo/ [accessed: 01 February 2010].

# Chapter 7
# Using Trajectory Data to Estimate Energy Consumption for Routing Purposes

Karin Kraschl-Hirschmann, Raphael Luz, Martin Fellendorf and Stefan Hausberger

## Introduction

In the European Union, the transport sector is responsible for generating the most greenhouse gases, and energy use in this sector is growing faster than in any other. Road transport is the largest energy consumer and accounts for approximately 83 per cent of the transport sector's energy consumption in the EU-25 countries in 2004 (excluding maritime transport and pipelines) (Eurostat 2011). This is equivalent to 290 million tonnes of oil, which represents over a quarter of the European Union's total final energy consumption, the latter of which includes the transport, industry, household and service sector.

Many approaches are being used to improve fuel efficiency and reduce emissions in the road transport sector. These include technical advancements in vehicles and fuel formulation, demand reduction, traffic controls, fleet optimization and measures designed to influence driving behaviour such as 'Eco-Driving' programs. Providing better information to drivers about energy use and emission using road shows, trainings, written information and websites have shown that up to a 15 per cent reduction in fuel consumption is possible (Onoda 2009).

Energy consumption is highly influenced by route choice. The fastest or shortest route is not necessarily the most ecological. Research has shown that different routes between the same origin and destination can often have very different results in terms of fuel consumption and emissions generation (Frey et al. 2008, Anh and Rakha 2008). For example, the influence of road gradients on energy consumption has been analysed in several studies (for example Park and Rakha 2006).

Most vehicle trip planners and navigation systems that are currently in use select the optimal path by minimizing distance, travel time or a combination of the two. In many cases, selecting a path based on the shortest distance will also minimize fuel consumption and emissions. However, there are cases where this is not true, particularly in highly congested areas and on routes with significant road gradients. Typically, fuel consumption and emissions increase with travel distance and speed. However, fuel consumption also increases at very low speeds, especially under stop-and-go conditions in congested networks.

Intelligent Transportation Systems (ITS), especially intelligent on-board navigation systems, could help to solve this problem by calculating not only the fastest or shortest routes, but more importantly the most energy efficient routes. Calculating the most energy efficient routes would require access to additional roadway link information; more specifically, information on the energy required to travel on each link under the current and predicted traffic conditions, road gradient information, road type and road surface conditions.

This chapter does not describe a new routing algorithm, but rather models and discusses the use of energy efficiency as a new objective function for providing drivers with routing recommendations. The method here uses link values based on energy consumption to identify the best paths. These link values represent the energy needed to overcome all possible driving resistance forces (which are, in turn, influenced by road characteristics, vehicle type and traffic conditions). The most innovative aspect of this method is its ability to generate these link values quickly enough to be suitable for real-time navigation applications. The link values are generated using functions that account for different driving modes, acceleration/deceleration rates differentiated by road type and traffic conditions. Instead of estimating fuel consumption, this method computes the actual energy required for each link and turning movement.

The following section presents a background on microscopic traffic flow and emission modelling, which was the basis for developing an integrated model to analyse different traffic management scenarios in former projects. For this work, the traffic flow model is used to generate a huge amount of datasets for analysing driving behaviour under different traffic conditions. Additionally, the emission model is used to validate the results of the developed weighing methodology. The section 'Energy consumption related to driving modes' presents the proposed methodology for estimating link-based energy consumption. In the section 'Validation of the developed methodology', a case study of the new method is described. The final section presents conclusions and recommendations for further research.

## Background

### *Emission model*

Several methods are available for calculating emissions from a given amount of traffic or individual driving trips. These methods range from statistical approaches to physical models that estimate vehicle emissions and fuel consumption. There are three basic types of physical models based on their level of detail. The three types are:

- Microscopic traffic flow simulation and instantaneous emission models;
- Traffic network and emission models (mesoscopic); and
- Inventory models (macroscopic).

Macro- and mesoscopic emission models such as COPERT or HBEFA (Handbook Emission Factors for Road Transport (Hausberger et al. 2009)) are used for high level studies such as national emission inventories or the modelling of large road networks where the analysis of single vehicles is usually not required.

PHEM (Passenger Car and Heavy Duty Emission Model (Hausberger 2003)) is a micro scale emission model that calculates the fuel consumption and emissions of individual vehicles based on vehicle longitudinal dynamics and emission maps. This model uses a backwards calculating approach (wheel to engine) as illustrated in Figure 7.1.

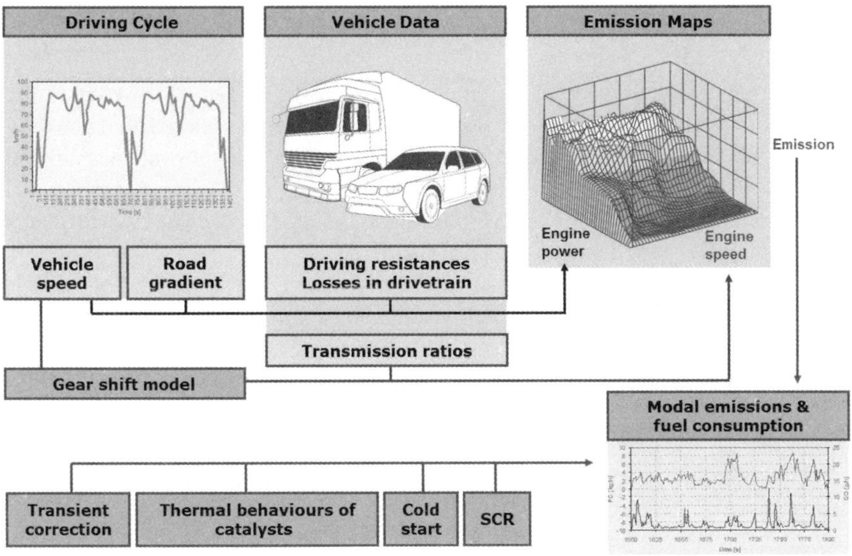


**Figure 7.1 PHEM micro scale emission model**

The PHEM model requires the following basic data:

- Vehicle data (for example, mass, driving resistances and gear ratios);
- Emissions map (that is, data describing emissions and fuel consumption over engine power and speed based on stationary and transient emission measurements from roller or engine test beds); and
- Driving cycle at 1 Hz resolution which defines vehicle speed, road gradient and optional information such as engine speed.

The PHEM model uses this data, along with vehicle speed, road gradient and driving resistances, to calculate the necessary propulsion power. The engine power also includes the drivetrain losses. The engine speed is determined by the vehicle speed, gear ratios and the actual gear. A gear shift model is used to provide realistic gear shift behaviour. Engine speed and power are used to interpolate emissions and fuel consumption using the appropriate emissions maps. PHEM also offers modules for considering the thermal behaviour of engines as well as catalyst or hybrid power train configurations.

PHEM includes a very extensive database containing vehicle data and emission maps. This data is compiled into average vehicle categories (for example, passenger cars, light duty and heavy duty vehicles with Otto and Diesel engines from EURO 0 to EURO 6) that are used to estimate traffic emissions on road networks. The same dataset was used for the calculation of the base emission factors of the HBEFA 3.1 for these vehicle categories.

*Combined traffic flow and emission model for environmental analyses*

Microscopic traffic flow simulation models allow users to analyse different traffic control strategies by modelling detailed vehicle movements within a road network. One big advantage of such models is that users can test different strategies without facing implementation costs. If the traffic is modelled at 1–10Hz and the model is well calibrated, the longitudinal movements of modelled vehicles will match reality adequately (Fellendorf and Vortisch 2010).

In most cases, the output measures of microscopic traffic simulation models are macroscopic traffic flow data such as aggregated values for volumes, densities, travel times and delay. Since the model recalculates every 0.1 to 1 second, the vehicle position along with current speed, distance to surrounding vehicles, current lane and anticipated route can be recorded with this precision as well. This means that the microscopic data output contains trajectory data for each individual vehicle including location and speed over time. Trajectory data can be matched against desired speeds, journey times for specific route sections and green times of relevant signal groups passed along the route of each individual vehicle. This provides detailed information on stopped and controlled delay.

In order to calculate the emissions caused by different traffic strategies at the granularity of an emission model like PHEM, a longitudinal trajectory with the attributes of the vehicle-driver-unit is needed. This means that traffic modellers need to define traffic flow parameters of VISSIM to obtain speed trajectories leading to realistic emissions. Several projects have been completed at the Graz University of Technology that investigate the potential of adaptive traffic signal control strategies regarding the reduction of emissions and fuel consumption (Hirschmann and Fellendorf 2009, Hirschmann et al. 2010). Results of these studies show that time resolution in the VISSIM simulation needs to be at least 1Hz to achieve realistic speed profiles for different driving characteristics. Thus, microscopic traffic flow simulators like VISSIM allow users to model different driver behaviour by using specific vehicle-driver parameter settings.

Hirschmann et al. (2009) proposed a toolbox where the microscopic traffic flow model is linked to the emissions model. The toolbox was used to identify an emission optimized traffic signal control strategy as part of a research project. Both models operate on single vehicle engine data with the same temporal resolution and level of detail. VISSIM generates trajectories for each vehicle within a specified link, sequence of links or sub-network. This trajectory data is recorded. PHEM then uses the trajectories to calculate the emissions of each vehicle over time.

PHEM can be combined with traffic models to calculate emissions for all vehicles defined in the traffic model's network. The vehicle and engine data needed to make this calculation are taken from the PHEM database on average vehicles. The vehicle category classification (cars, LCV, HDV, buses, and so on) is taken from the traffic model results. Classification into fuel type, weight and EURO classes are chosen automatically from PHEM for each single vehicle according to a predefined fleet composition. The fleet composition must be set based on local vehicle registration data. Due to the continuous renewal of the vehicle fleet, it must be revised each year.

PHEM uses trajectory data produced by VISSIM in a post process. The VISSIM data contains time, speed, location and vehicle type of all individually simulated vehicles. If information about the simulated road network is available, PHEM can assign the calculated emissions to road segments as input for emission dispersion models.

When coupling traffic and emission models, additional important impacts such as gear shift behaviour and cold start effects must be considered. In this research, the gear-shift model embedded within PHEM (based on Zallinger (2010)) was used.

To perform the simulation as realistically as possible, default values including driving behaviour and acceleration rates must be adjusted. Calibration is usually limited in other microscopic traffic flow studies to traffic volumes, delays, queue lengths, travel times and travel speeds.

A very important factor for calculating emission levels is acceleration behaviour. Individual acceleration rates can be recorded very accurately using GPS-drives. In this research, several different acceleration/deceleration distributions were developed to simulate driving patterns as realistically as possible. These acceleration distributions closely matched the observed and measured acceleration rates.

When VISSIM is well-calibrated and the lane allocation application is adjusted correctly, realistic vehicle trajectories are provided. It is assumed that the majority of calibration efforts completed in this research can also be used in other urban applications.

The toolbox prototype has proven to be a valuable framework for preparing detailed emissions calculations that fit very well with measured single vehicle emissions data. Figure 7.2 shows the match of test trips and VISSIM simulation for the City Center of Graz.

As shown in Figure 7.2, the VISSIM simulation fits quite well with the test trips and only a few test trips with a very high journey time were not covered by the simulation.

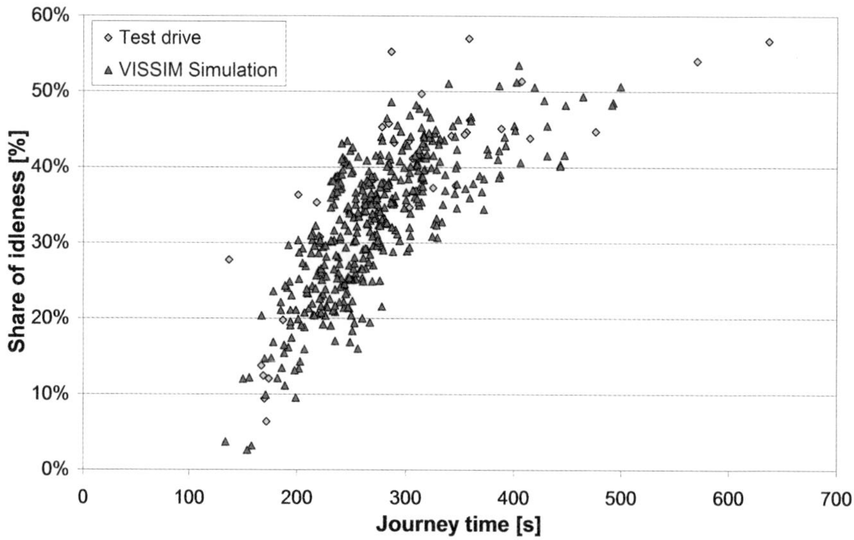


**Figure 7.2 Relative share of idling for urban vehicle trips with empirical and simulated data**

*Traffic flow trajectories for analysing driving patterns*

Due to the limitations of computing power, it is not possible to model fuel and energy consumption in large networks using the detailed microscopic approach described in the previous section. Therefore, in earlier research, some authors proposed an approach for calculating emissions and fuel consumption by averaging multiple individual speed trajectories (Kraschl-Hirschmann et al. 2011). This method showed promising results and therefore additional scenarios were tested that included the use of specific values for every link. The results support the idea of developing a method for estimating the required energy for each link by using the average current link speed and an additional indicator describing speed variability on the link (for example, level-of-service (LOS) or volume-to-capacity (v/c) ratio).

This approach requires collecting an extensive amount of data to analyse driving patterns for different road types and traffic conditions (represented by the v/c-ratios). The collection of real world GPS-measurements is very expensive and matching this data with actual traffic counts is very complicated. The previous sections have shown that a microscopic traffic model could be used to produce data describing driving cycles and that this data could be used as input for an emissions model with good results.

In order to calculate emissions during a trip, data about different driving modes (cruising, idling, accelerating and decelerating) as well as acceleration and deceleration rates are important. This means that the parameters needed to analyse emissions are the same as those needed to calculate energy consumption. In this research, test areas with different road types and traffic influences were modelled to produce driving trajectories for any possible traffic-road type combination. The test areas modelled in the research were calibrated using previously developed acceleration and deceleration functions as these have the most influence on energy relevant driving behaviour parameters. To calculate the emission factors for HBEFA 3.1 (Handbook Emission Factors of Road Transport), driving cycles (raw data) were used. The emission factors are provided for different LOS, road types and roadway gradients and raw data is disaggregated in the same way.

## Energy consumption related to driving modes

### *General information*

Specific link information about driving behaviour is needed to analyse the energy consumption of a trip. A vehicle trip can be described by four different driving modes: acceleration, deceleration, cruising and idling. The European Urban Driving Cycle contains fixed periods of acceleration, idling, cruising and deceleration designed to represent driving in typical European capitals. Similarly, the Extra Urban Driving Cycle (EUDC) is designed to account for more aggressive, high speed driving modes (although speeds never exceed 120 km/h). Real vehicle trips and idealistic driving cycles alike are characterized in terms of the four distinct driving modes. Since analysis of driving cycles requires speed and acceleration data over time, the driving modes are defined in terms of time as follows:

1. Cruising time occurs at driving conditions with very little acceleration or deceleration. Since it is nearly impossible to keep exactly constant speed under realistic conditions, the acceleration may fluctuate within narrow boundaries ($-0.125$ m/s$^2$ and $+0.125$ m/s$^2$).
2. Idle time consists of the time when the vehicle is stopped (speed = 0 m/s) but engine or auxiliary engine parts are still operating.
3. Deceleration time is the period where speed decreases at a rate smaller than $-0.125$ m/s$^2$.
4. Acceleration time is the period where speed increases at a rate larger than 0.125 m/s$^2$.

Every vehicle trip contains several of these four modes, although the duration and magnitude spent in each of the modes may differ significantly among trips. For example, the idling time will be relatively short during interurban trips on uncongested routes while it may be long for trips in congested urban centres. The share of time spent in each mode has a significant impact on fuel consumption and emissions.

A motor vehicle engine provides motive force to overcome resistance and move the vehicle forward. According to the conservation of energy law, the total amount of energy in an isolated system remains constant over time. Therefore, the sum of all resistances experienced during a trip will be equal to the energy required to complete the trip. The resistances relevant to driving are: rolling resistance, air resistance, acceleration resistance, deceleration resistance, gradient resistance and auxiliary system resistance.

Each driving mode contains a different set of resistances. In cruising phases acceleration and deceleration are not relevant while on the other hand rolling, air, gradient and auxiliary resistances are present. During acceleration and deceleration, the change of vehicle speed over time has a significant influence on energy consumption. Table 7.1 summarizes the types of energy needed to overcome each resistance depending on driving mode.

**Table 7.1 Different driving modes of a trip and the occurring driving resistances**

| Time share | $E_{roll}$ | $E_{gradient}$ | $E_{air}$ | $E_{dec}$ | $E_{acc}$ | $E_{aux}$ |
|---|---|---|---|---|---|---|
| Cruising time | ✓ | ✓ | ✓ | | | ✓ |
| Idling time | | | | | | ✓ |
| Decelerating time | ✓ | ✓ | ✓ | ✓ | | ✓ |
| Accelerating time | ✓ | ✓ | ✓ | | ✓ | ✓ |

$E_{roll}$, $E_{air}$ … Energy needed to overcome rolling/air resistance
$E_{gradient}$ … Energy needed to overcome the resistance due to the road gradient
$E_{acc}$, $E_{dec}$ … Energy required/recovered due to inertia during acceleration/deceleration
$E_{aux}$ … Energy demand of auxiliaries

For real applications, the only information available about current traffic conditions is the average speed that can be driven under the current traffic conditions and LOS (or the v/c-ratio) calculated using actual traffic data. Typically, the speed information for every link in a traditional navigation system is static and not adapted or changed based on traffic conditions. Also, roadway gradient and road type information could increase the benefit as additional static information.

For online purposes, a fast way to define the time spent in each of the four different driving modes on each link should be possible with the described input parameters and for every type of traffic condition and road type. Also, different functions for estimating the average acceleration/deceleration over the average speed should be available (differentiated by LOS, road type and road gradient).

The general approach proposed in this research to calculate energy consumption for each link is illustrated in Figure 7.3. The method starts with real-time traffic. It uses this information to estimate the required energy based on a time share function (duration of each driving cycle phase) and an acceleration/deceleration function. Both functions are customized for each road type based on different traffic conditions. The following sections describe the method in more detail.

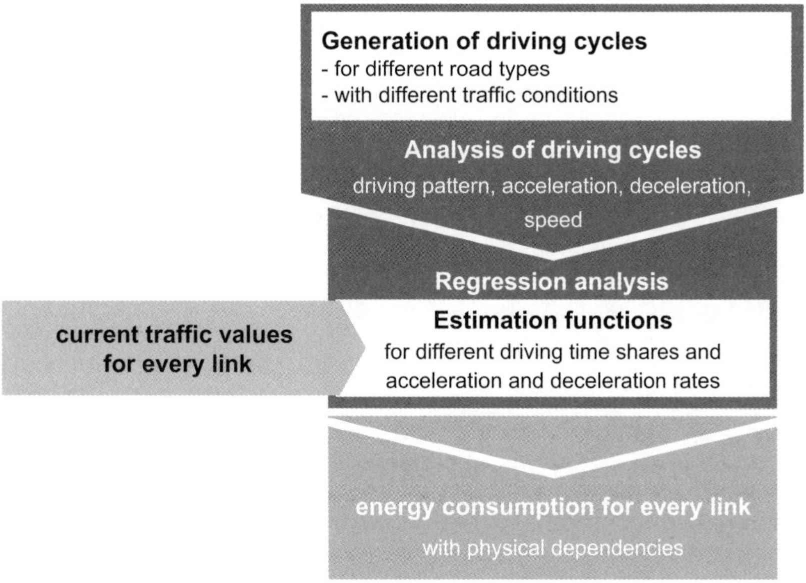


**Figure 7.3 System architecture to estimate the energy consumption of links**

So far only v/c-ratio and average link speed are taken as input data for each link. These link-based volume and speed data are used to define the share of time spent in each of the different driving modes (cruising, idling, accelerating and decelerating) as well as the values for the average acceleration and deceleration rates. The energy needed can be then calculated using the average positive acceleration.

This research calculates energy consumption at the wheels, which means that losses in transmission and drivetrain are not considered. The types of energy needed to overcome the different resistances experienced during each driving mode are summarized in Table 7.1 and Equation (7.1).

$$E = \begin{Bmatrix} E_{roll} + E_{air} + E_{grade} + E_{aux} \\ E_{aux} \\ E_{roll} + E_{air} + E_{grade} + E_{dec} + E_{aux} \\ E_{roll} + E_{air} + E_{grade} + E_{acc} + E_{aux} \end{Bmatrix} * \begin{Bmatrix} \text{temporal share}[\%] \text{ of cruising mode} \\ \text{temporal share}[\%] \text{ of idling mode} \\ \text{temporal share}[\%] \text{ of deceleration mode} \\ \text{temporal share}[\%] \text{ of acceleration mode} \end{Bmatrix} \tag{7.1}$$

Assuming there are no transmission losses, the energy required to overcome each of the different resistances can be calculated using Equations (7.2a) to (7.2e).

Equation (7.2a) can be used to calculate the energy needed to overcome rolling resistance.

$$E_{roll} = m*g*\left(F_{R0} + F_{R1}*v\right) \qquad (7.2a)$$

with: $E_{roll}$ rolling resistance [J]
m vehicle mass (including loading) [kg]
g gravitational acceleration [m/s$^2$]
v vehicle speed [m/s]
$F_{R0}$, $F_{R1}$ coefficients of the rolling resistance

The rolling resistance coefficients $F_{R0}$ and $F_{R1}$ are determined by coast-down tests carried out by the manufacturer.

Equation (7.2b) can be used to calculate the energy needed to overcome air resistance.

$$E_{air} = c_w*\frac{\rho}{2}*A*v^2 \qquad (7.2b)$$

with: $E_{air}$ air resistance [J]
$c_w$ drag coefficient [-]
A frontal area of the vehicle [m$^2$]
ρ density of the air [kg/m$^3$]

The values for $c_w$ and the frontal area of the vehicle are given by the vehicle manufacturer.

Equation (7.2c) can be used to calculate the energy needed to overcome gradient resistance.

$$E_{grade} = m*g*gradient*v \qquad (7.2c)$$

with: $E_{grade}$ gradient resistance [J]
m vehicle mass (including loading) [kg]
gradient road gradient [/100]
v vehicle speed [m/s]

Equation (7.2d) can be used to calculate the energy needed by the vehicle to accelerate and decelerate.

$$E_{acc/dec} = m*a*v \qquad (7.2d)$$

with: $E_{acc/dec}$ acceleration resistance [J]
m vehicle mass (including loading) [kg]
a vehicle acceleration [m/s$^2$]
v vehicle speed [m/s]

The energy demand of auxiliaries consists of energy not directly used for the vehicle's motion, for example powering the compressor for air conditioning and the hydraulic pump for power steering. The auxiliaries are considered together and their energy use is assumed to be constant over time, or in other words, independent of speed and acceleration. Equation (7.2e) can be used to calculate the energy needed to provide vehicle auxiliary services.

$$E_{aux} = const \qquad (7.2e)$$

with: $E_{aux}$ auxiliary energy demand [J]

*Information requirements*

The proposed methodology uses several types of information. Much of this information exists but is not commonly used in traditional navigation systems. At best, link-based average speed and LOS data are currently availablc information in a navigation system network. The average speed should be the speed that can be achieved by 50 per cent to 85 per cent of all vehicles. Until recently, link-based speed information in traditional navigation systems was static and not adapted to actual traffic and road conditions. Real-time speed data is a significant improvement. This navigation data can be further improved by adding information about link gradient, capacity and road type. This information can be added easily if road authorities release the appropriate data.

The proposed method assumes v/c-ratio and current speed for each link as input values. The v/c-ratio will be calculated using a travel demand and assignment model. Current values for traffic amount and speed are associated with each link. Next, the time shares spent in each driving mode (cruising, idling, accelerating and decelerating) as well as the values for average acceleration and deceleration rate are determined. The average acceleration rate can be used to calculate the needed energy using Equation (7.2d). The different energy components for each driving phase are discussed above. It should be noted that weather information such as wind direction (influences acceleration), temperature (influences use of air conditioning) and precipitation (influences rolling resistance) could help to improve the accuracy of the required energy calculations. However, these additional influences have not yet been incorporated into the methodology.

*Functions for the different driving shares*

On any given link the amount of time spent in each driving mode depends on traffic conditions, road gradient and road type, among other factors. For example, the share of time spent in the acceleration and deceleration mode will increase when the traffic state changes from free-flow to stop-and-go conditions. Similarly, cruising on a level road requires less energy than cruising on an uphill gradient.

An efficient way to determine driving mode time shares is to use predefined estimation functions for each road type. Producing these estimation functions requires a reliable data source. Figure 7.4 shows an example of the estimation functions for an average Euro 4 vehicle with a road gradient of 0 per cent and a specific road type. In this figure, every data point is a specific driving mode time share over the average speed for an individually simulated driving cycle. These data points are then used to construct a best fit curve using regression.

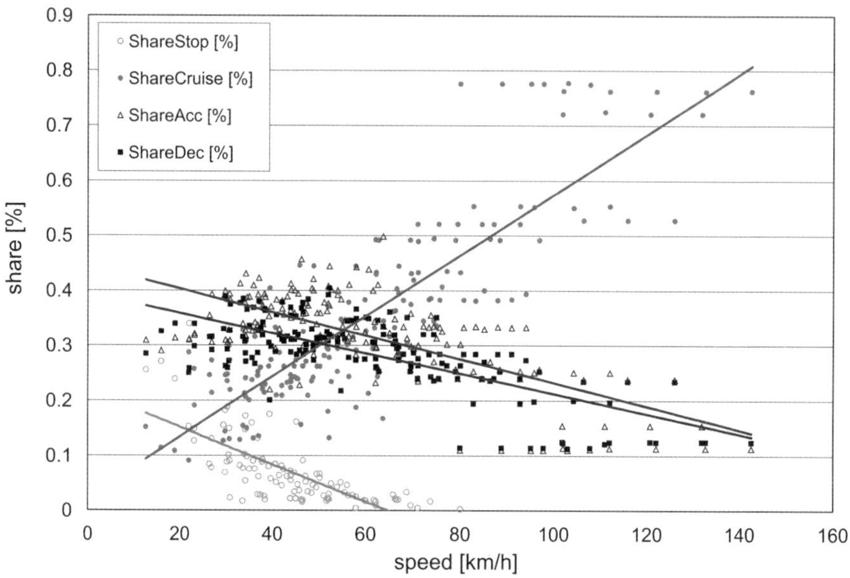


**Figure 7.4 Example of driving mode time share versus average link speed created using HBEFA driving cycles ($R^2$=0.6 – 0.75)**

*Functions for acceleration/deceleration profile*

The acceleration and deceleration patterns during a driving cycle depend on traffic conditions. Acceleration rates may differ for the same link even at the same average speed. Therefore, the average link speed and its associated acceleration and deceleration values are needed to calculate the driving forces and to estimate the required engine power. To obtain this information, an estimation function for the acceleration and deceleration rates of different traffic states, road types and roadway gradients was developed using the raw data from the HBEFA. Figure 7.5 shows an example of the estimation functions calculated for an average Euro 4 vehicle with a road gradient of 0 per cent and urban road type. These data were used to construct a best fit curve and determine the resulting equation and the line-fit parameters for the acceleration and deceleration values.

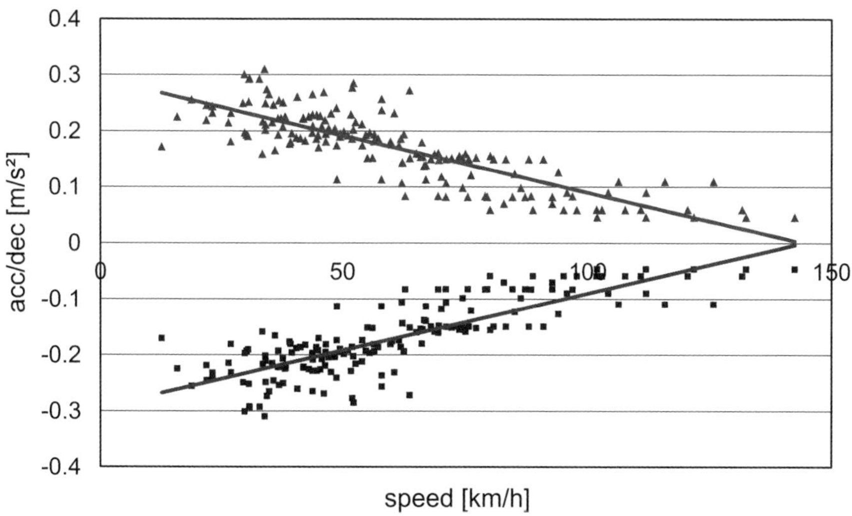


**Figure 7.5 Example of acceleration and deceleration rates versus average link speed created using HBEFA driving cycles ($R^2$=0.7)**

*Energy calculation using the estimation functions*

The final step in estimating energy consumption on each link consists of multiplying the link-based driving mode time shares by the energy required per driving mode. The energy required for each driving mode is estimated using the energy equations. This is shown in Equation (7.1) where E represents the energy required to overcome the driving resistance on a specific link based on the current traffic state, road type, road gradient and vehicle type.

The actual fuel/energy consumption can be calculated by characteristic curves for the engine efficiency as a function of the engine power output. In this research characteristic curves were computed by using the simulation results of the PHEM model.

## Validation of the developed methodology

The instantaneous emission model PHEM was taken as a reference tool to estimate power requirements of individual trips. PHEM is widely accepted in Europe as a standard for microscale emission modelling due to its vast amount of continuous field measurements and vehicle emission tests on chassis dynamometers (Hirschmann et al. 2010) which serves to keep the engine maps up-to-date with actual vehicle fleets.

In this research we compared the results of the required engine power from the instantaneous emission model PHEM to the results of the energy estimation using our proposed approach (converted into power).

In our proposed approach, the vehicle longitudinal dynamics model calculates the engine power output and speed from physical interrelationships. This means that any driving condition can be analysed using our approach.

As part of this research we applied the energy estimation algorithm for different road types (described in the previous sections) to a real test site. The test area included about 14 km of the A7 motorway north of Linz, Austria. Since the motorway is within the urban area, the speed limit changes from 130 to 100 and reduces to 80 km/h. The test site was divided into 8 segments with lengths ranging between 600 and 3,300 metres.

Over 900 test trips using GPS instrumented vehicles were made to collect the case study data. For validation purposes, the GPS trajectories measured in the field were supplemented by point data on speed and volume. Additionally, journey times were recorded using automatic number plate recognition.

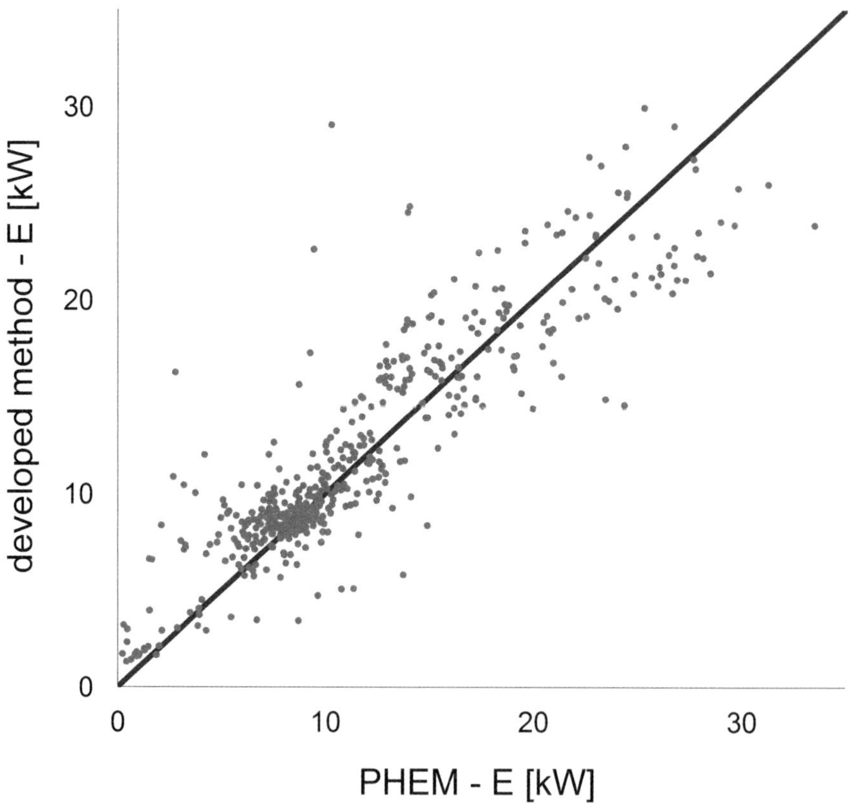


**Figure 7.6 Comparison of methods for calculating engine power demand ($R^2$=0.8)**

The estimation functions for the different drive mode time shares and acceleration/deceleration rates were created with the HBEFA cycles, using the characteristics of a diesel EURO 4 vehicle. For each GPS trajectory, the required power was computed using PHEM and the proposed method. Figure 7.6 plots the results of both methods. As shown, there is a very good fit between the detailed calculation results from PHEM and the proposed method.

## Conclusions

Link-based energy consumption data are not only important for optimal routing within navigation systems; they are also relevant for the automotive sector to consider for in-vehicle components such as automatic speed control systems. Energy consumption data can also be used by electric vehicles (EV) to, among other functions, help optimize battery recharging processes and reduce energy consumption.

Vehicle energy consumption is directly related to driving mode (acceleration, deceleration, cruising and idling). The duration of each driving mode depends on driver behaviour and traffic conditions. In order to estimate energy consumption most accurately, these driving modes should be known either ex-post or a-priori. By taking advantage of current developments in communication between vehicles and road infrastructure, it will be possible to obtain all the energy relevant data and provide it to the driver. Further, by using real-time information from a macroscopic traffic model, the algorithm proposed in this research provides the data necessary to identify routes which minimize energy consumption. The calculation of energy consumption will utilize network data such as road gradient and road type, which is currently available in geographic information systems (GIS) for many cities and infrastructure owners. By connecting this data with the proposed energy consumption algorithm, link-based energy consumption can be calculated quite accurately. Information on actual weather and road conditions could further improve accuracy.

The proposed method has been tested and found to generate similar results to more complex methods. The proposed method can estimate power requirements relatively quickly, an important aspect needed for navigation and other real-time forecasting purposes. The proposed method considers all driving resistances at the wheel (that is, without drive train losses) and considers auxiliary power demand as a constant average amount. The proposed approach could be improved though to consider real-time weather information, which could be especially important for EVs as auxiliary power consumers, such as climate control, can have a significant influence on energy consumption and therefore driving range.

**Acknowledgements**

The authors acknowledge the support provided by the city councils of Graz and Linz for the GIS-based road network of their cities.

**References**

Anh, K. and Rakha, H. 2008. The effects of route choice decisions on vehicle energy consumption and emissions. *Transportation Research Part D: Transport and Environment*, 13 (3), 151–67.

Eurostat, 2011. *Europe in Figures, Eurostat Yearbook 2011*. European Union.

Fellendorf, M. and Vortisch, P. 2010. Microscopic Traffic Flow Simulator VISSIM, in *Fundamentals of Traffic Simulation*, edited by J. Barcélo. New York: Springer, 63–93.

Frey, H.C., Zhang, K. and Rouphail, N.M. 2008. Fuel use and emissions comparison for alternative routes, time of day, road grade, and vehicle based on in-use measurements. *Environmental Science & Technology*, 42 (7), 2483–9.

Hausberger, S. 2003. *Simulation of Real World Vehicle Exhaust Emissions*. Graz: Verlag der Technischen Universität Graz.

Hausberger, S., Rexeis, M., Zallinger, M. and Luz, R. 2009. *Emission Factors from the Model PHEM for the HBEFA*. Version 3. Report Nr. I-20/2009 Haus-Em 33/08/679.

Hirschmann, K. and Fellendorf, M. 2009. *Emission minimizing traffic control – simulation and measurements*. Proceedings of the mobil.TUM 2009, Munich, Germany.

Hirschmann, K., Zallinger, M., Fellendorf, M. and Hausberger, S. 2009. *GAVe – neue Regelungen – weniger Schadstoffe*. 7. Sommerakademie. Graz.

Hirschmann, K., Zallinger, M., Fellendorf, M. and Hausberger, S. 2010. *A new method to calculate emissions with simulated traffic conditions*. Proceedings of the IEEE Intelligent Transportation Systems Conference 2010, Madeira Island, Portugal.

Kraschl-Hirschmann, K., Zallinger, M., Luz, R., Fellendorf, M. and Hausberger, S. 2011. *A method for emission estimation for microscopic traffic flow simulation*. Proceedings of the FISTS Conference 2011, Vienna, Austria.

Onoda, T. 2009. IEA policies – G8 recommendations and an afterwards. *Energy Policy*, 37 (10), 3823–31.

Park, S. and Rakha, H. 2006. Energy and environmental impacts of roadway grades. *Transportation Research Record* (1987), 148–60.

Zallinger, M. 2010. *Microscopic Simulation of passenger car emissions*. Doctoral Thesis Technical University Graz. Graz.

Chapter 8

# Consumer Response to Accumulated Transport Charges: Developing Single Attribute Mixture-Amount Preference Experiments

Elaheh Khademi, Gamze Dane, Harry Timmermans and Dujuan Yang

## Introduction

Many academics and transportation planners seem convinced that pricing policies may be one of the most effective policy instruments to induce travellers to choose different routes, departure times and transport modes, thus minimizing congestion, emissions or optimizing system use otherwise. The latter objectives are closely associated with concepts of sustainable mobility in urban areas with respect to social equity, economic efficiency and environmental responsibility. Not surprisingly, therefore, different kinds of pricing strategies have been studied and put into practice in various cities and countries. For decades, tolls have been charged on certain roads with the goal of not only stimulating particular behaviour, but also of recuperating building and maintenance costs. In addition, various road and congestion pricing strategies have been discussed at length in countries such as the Netherlands, and such schemes have been implemented in some cities elsewhere in the world. More recently, parking charges have become popular across the world, not only in commercially managed parking garages and public inner-city parking spaces, but also in and around shopping centres located in neighbourhoods and peripheral areas.

The impact of pricing policies has been studied at length, especially in the context of congestion pricing and toll roads. For example, Arentze, Timmermans and Hofman (2004) found that for work trips, route change is the most important adaptation of people under congestion pricing, followed by departure time adjustment. In the case of non-work trips, the willingness to adapt is smaller and changing the route or switching to bike travel is the dominant response. Similarly, Keuleers et al. (2005) investigated the effect of a hypothetical road-user charging scheme on travel behaviour, activity participation and rescheduling patterns in the city of Newcastle upon Tyne in England. The results showed that the scheme was effective in reducing car use during peak times in the city but that overall activity participation remained largely unchanged. Yan, Small and Sullivan (2002) studied the effect of value pricing, where the traveller is offered a choice of price and service quality in Orange County, California. Results here confirmed that the operators in both value pricing experiments in California were able to fine tune tolls in order to keep these lanes busy while maintaining congestion-free conditions. Others also found that shifts to different vehicle occupancies or times of day in response to toll changes were very small because of the existence of the option to change route (see Burris and Pendyala 2002, Peters et al. 2011, Bhat and Castelar 2002).

As for toll roads, Dissanayake and Kouli (2007) investigated driver response to a proposed toll motorway project connecting the cities of Corinth and Patras in Greece. The overall objective of this research project was to develop discrete choice models to investigate driver behaviour in regards to inter-city route choices and to explore driver attitudes to road pricing. The results showed that drivers prefer the new toll motorway over the existing alternative routes. In their model, both travel time and travel cost coefficients were negative and highly significant, indicating that travel utility decreases with increasing travel time and cost.

Compared to congestion pricing and toll roads, less is known about the effects of parking charges. A study reported by Danwen, Wei and Shihui (2010) focused on the effects of parking charges on travel behaviour. Using elasticity analysis, the paper concluded that residents' travel modes change with increasing parking rates. It also revealed the proportion variation between car and bus transit for different parking rates and analysed the elasticity of parking rates and bus fares. Guan et al. (2005) also assessed the effects of parking policies in the form of meters, discounted transit passes, and other transportation demand management (TDM) strategies on travel and parking behaviour. Emphasis here was made on the relationship between parking pricing and mode choice. It was shown that the installation of the meters and the discounted transit pass program had a significant effect on decreasing the drive-alone mode.

This brief review suggests that traveller adaptive behaviour to various transport charges has been the subject of a considerable amount of empirical work. As a result, our understanding of the impact of transport charges has increased substantially in recent years. Nevertheless, several aspects of transport charges have still not been studied in much detail. A key question remaining concerns the accumulation of possible charges. To date, most studies have typically addressed only a single charge. However, consumer response to an instrument such as parking policy may depend on other charges. Moreover, travellers face the problem on how to allocate their available budget to alternative transport charges or other expenditures.

In order to address this question, this chapter describes an experimental approach to better understand how travellers respond to such accumulated pricing schemes, subject to budget constraints. More specifically, it allows conclusions concerning questions such as (i) how do travellers allocate part of their travel budget to different types of pricing policies, (ii) to what extent does this allocation depend on the total available budget, and (iii) how do they trade-off the budget allocation and the gain experienced by choosing the priced choice alternative. This approach should allow researchers to assess the effects of pricing policies on travel and therefore sustainability.

This chapter is organized as follows. First, the basic principles of mixture amount models and how they can be applied to the research issue at hand will be discussed. As part of this section, the case will be elaborated, including the trade-off between budget allocation and varying gains. After, operational decisions in the design of the experiment and the sample will be covered. The chapter will end with a summary and some conclusions.

## Single attribute mixture-amount models

Stated preference models are commonly used in transportation research. They systematically vary the levels of a set of attributes, according to an orthogonal fractional factorial design, and ask respondents to rate the resulting attribute profiles on some preference scale. When the purpose of the study is to estimate a choice model, these attribute profiles are placed into choice sets and respondents are asked to choose the choice alternative they like best. The design strategy to create these choice sets depends on the choice model and the specific purpose of the analysis.

Typically, respondents are requested to choose the option they like best. Choices are usually not limited to satisfy any budget constraint. If the choice model should however satisfy budget constraints, this complicates the design of the experiment as the choice percentages must sum to 100 per cent. When the attribute levels are subject to the constraint of summing to one (or 100 per cent), standard mixture designs can be used that fit to standard models such as Simplex-Lattice and Simplex-Centroid. When mixture components are subject to additional constraints, such as a maximum and/or minimum value for each component, designs other than the standard mixture designs (such as constrained mixture or Extreme-Vertices) are appropriate. The mixture components are non-negative proportionate amounts of the mixture. It is also possible that the allocation only concerns a fraction of the budget. In this case, the variable proportions can be rewritten as scaled fractions so that the scaled fractions sum to one.

A mixture consists of q linearly independent variables that are constrained to a $(q-1)$-dimensional dependent space. The levels of the q variables sum to a constant, which equals the total amount. Figure 8.1 gives an illustration, showing three linearly independent variables; $x_1$, $x_2$, and $x_3$, with standardized ranges from 0 to +1.

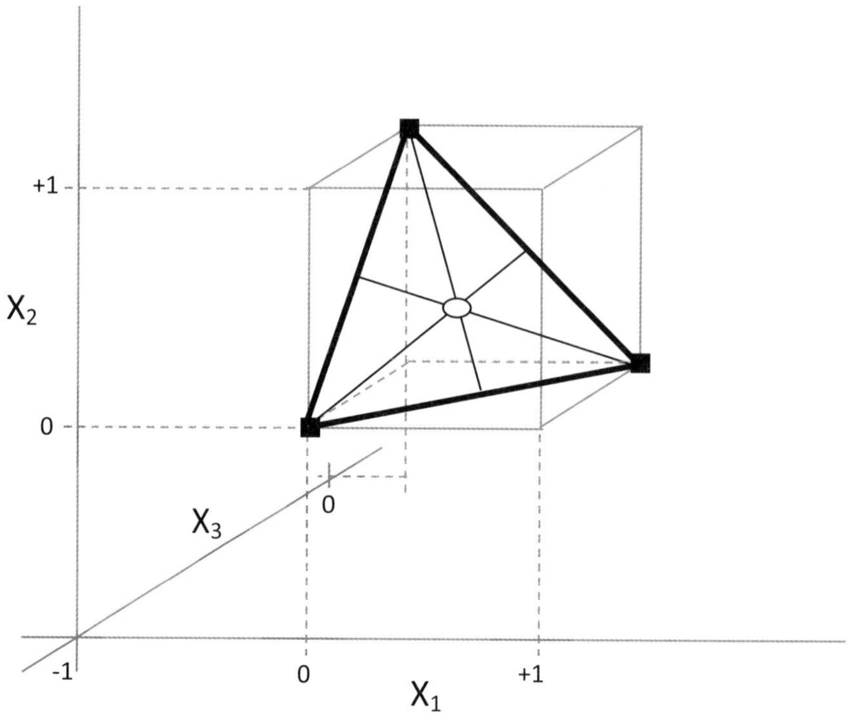


**Figure 8.1 Mixture space for three variables**[1]

Within the standardized independent space, the ($q$-1)-dimensional constrained space is represented by the triangle within the cube. This space is called a simplex space. The mixture constraint is presented in equation form below.

$$\sum_{i=1}^{q} x_i = 1 \qquad (1)$$

Figure 8.2 shows the ($q$-1)-dimensional simplex space for three experimental variables. All points on the surface or in the interior of the simplex sum to one. It shows the typical simplex coordinates of the three vertices; (1, 0, 0), (0, 1, 0), and (0, 0, 1), and the centre point; with coordinates (1/3, 1/3, 1/3). In addition, it shows the three edge midpoints; with coordinates (1/2, 1/2, 0), (1/2, 0, 1/2), and (0, 1/2, 1/2), and three interior points; with coordinates (2/3, 1/6, 1/6), (1/6, 2/3, 1/6), and (1/6, 1/6, 2/3).

1 See www.qualitycoach.net/doefusionmix.htm [accessed: 30 November 2012].

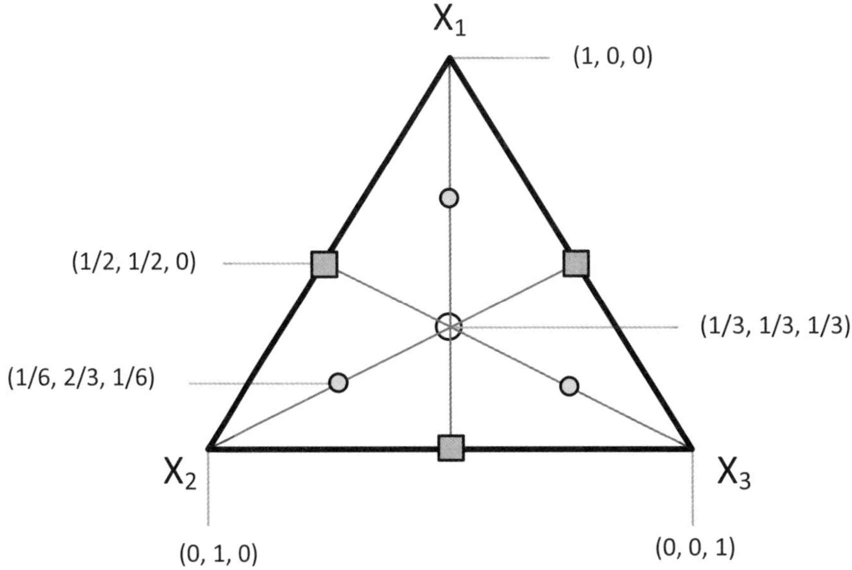


**Figure 8.2 Simplex space for three experimental variables**[2]

As mentioned above, Simplex-Lattice and Simplex-Centroid are two designs used to provide design points in mixture experiment. Scheffe (1958) introduced simplex-lattice designs for mixture experiments and developed polynomial models which have exactly the same number of terms as there are points in the associated designs. The designs, consisting of a symmetrical arrangement of points, are referred to as $\{q, m\}$-lattices. The models, expressed in canonical form by Scheffe (1958), are characterized by their relatively simple form and are thus easy to use for predicting the response over the simplex factor space. Corresponding to the points in a $\{q, m\}$-lattice, the proportions used for each of the q components have the m + 1 equally spaced values from 0 to 1, that is, $x_i = 0, 1/m, 2/m, \ldots, 1$, and all possible mixtures with these proportions for each component are used. The number of points in a $\{q, m\}$-lattice is $\binom{m+q-1}{m}$ and the lattice designs are characterized by their simplicity of construction (Scheffe 1958).

With a simplex-Centroid design, the design points correspond to $q$ permutations of (1, 0, 0, …, 0), $\binom{q}{2}$ permutations of (.5, .5, 0, …, 0), $\binom{q}{3}$ permutations of (1/3, 1/3, 1/3, 0, …, 0), …, and finally the single ($1/q$, $1/q$, …, $1/q$) mixture. The number of distinct points is $2^q - 1$ (Raghavarao, Wiley and Chitturi 2011).

In mixture experiments, the measured response is assumed to depend only on the relative proportions of the components in the mixture and not on the amount of the mixture. However, the amount of the mixture (in other words, the total available budget) can also be varied as an additional factor in the experiment.

2 See www.qualitycoach.net/doefusionmix.htm [accessed: 30 November 2012]

These are so-called mixture-amount experiments. In fact, a this type of experiment describes the relationship between the values of one or more response variables and allocations of available budgets or resources to a group of attributes. It also finds the optimal allocation of the available budget.

In a cost-based mixture approach, alternatives consist of profiles in which the costs of attributes satisfy the inequalities and the combined cost of the attributes is the same for all profiles. That is,

$$\sum_{i=1}^{q} c_i / U = 1$$

where $c_i$ is the cost of attribute $i$, $U$ is the total amount to be spent across all attributes, and $c_i / U = x_i$. According to the simplex-lattice design, the canonical form of the first-order mixture model is defined:

$$E(y) = \sum_{i=1}^{q} \beta_i x_i \quad (2)$$

Equation 3, also, shows the second-degree model:

$$E(y) = \sum_{i=1}^{q} \beta_i x_i + \sum_{\substack{i,j=1 \\ i<j}}^{q} \beta_{ij} x_i x_j \quad (3)$$

where E(y) is the response of a mixture (x1, x2 … xq ). It may be noted that not all simplex mixtures of {q, m} are needed to estimate the parameters of a second-degree canonical model. As total amounts and their allocation can affect the response, different levels of total amount can be considered in a mixture-amount experiment. This means that for the total amount, U, r different levels indicated by u1, u2, …, ur (r≥2) may be considered (Raghavarao and Wiley 2009).

In addition to different levels of amount, attributes may also have different levels. This point, to the best of our knowledge, has not yet received attention in mixture-amount experiments. To illustrate attributes, when selecting between transportation modes for a trip to work, consider three available modes as our attributes: car, bus and train. Each of these is different in terms of safety, comfort, travel time and so on. When people choose to trade-off across attributes, they may consider the amount spent as well as the possible benefits (safety, comfort, travel time, and so on) they gain by choosing each attribute. Accordingly, the allocation for any amount may be influenced by varying attributes levels. We address this issue here by attaching the levels to the attributes and as a result, go beyond a classical mixture-amount design. It should be noted that attached levels of attributes bring additional utility to the mixture model. Therefore, we measure the non-linear effects of attribute levels using effect coding.

## Experimental design

The design of a mixture experiment typically involves the following steps (Piepel and Cornell 1994):

*Define the objectives of the experiment*

Formally, the choice problem addressed in the previous section is a combination of (i) a budget allocation task, repeated for (ii) multiple amounts in which (iii) the allocations for any amount may be influenced by varying attributes levels.

*Select the mixture components and any other factors to be studied*

In the present study, we assume a mixture-amount experiment with three attributes as pricing options: toll road, congestion pricing and parking price, with three different budgets, $u_1$, $u_2$, and $u_3$, to be allocated. The mixture of the attributes is denoted by $x_1$ (toll road), $x_2$ (congestion pricing) and $x_3$ (parking price), ($0 \leq x_1$, $x_2$, $x_3 \leq 1$ and $x_1+x_2+x_3=1$). Moreover, each attribute has a combined level as time benefits are related to the pricing strategies.

The simplex lattice design {3, 3} creates 10 mixtures and the second-degree canonical model has 6 parameters. We consider three amounts of budgets to be allocated to attributes and combine the mixtures with amounts to form the choice sets of size 3, using Balanced Incomplete Block (BIB) design with seven symbols. When the number of replications of all pairs of treatments in a design is the same, we have an important class of designs called Balanced Incomplete Block designs (Colbourn and Dinitz 2007, Stinson 2003). Thus a BIB design, an arrangement of $v$ treatments in $b$ blocks each of size $k$ ($k < v$) is such that:

1. Each treatment occurs almost once in a block;
2. Each treatment occurs in exactly r blocks and
3. Each pair of treatments occurs together in exactly $\lambda$ blocks.

Table 8.1 below shows seven mixtures of three attribute mixture design.

**Table 8.1 Three attribute mixture design**

| Mixture | Attribute 1 | Attribute 2 | Attribute 3 |
|---|---|---|---|
| 1 | 1 | 0 | 0 |
| 2 | 0 | 1 | 0 |
| 3 | 0 | 0 | 1 |
| 4 | 2/3 | 1/3 | 0 |
| 5 | 0 | 2/3 | 1/3 |
| 6 | 1/3 | 0 | 2/3 |
| 7 | 1/3 | 1/3 | 1/3 |

Considering three amounts, a mixture design of 7 is used three times of each allocation which creates 21 profiles. With v = 7 = b, r = k = 3, l = 1, 21 profiles are organized into 7 choice sets of 3 profiles each (plus a no-choice option in each set). The 7 choice sets with these 21 profiles are given below.

**Table 8.2 Choice sets for three-attribute mixture design with three amounts (ui(t) denoted the t-th mixture with amount ui,t=1,2,...7; i=1, 2, 3)**

| Choice Set | Profile a | Profile b | Profile c | Base Alternative |
|---|---|---|---|---|
| 1 | $u_1(1)$ | $u_2(2)$ | $u_3(4)$ | No Choice |
| 2 | $u_1(2)$ | $u_2(3)$ | $u_3(5)$ | No Choice |
| 3 | $u_1(3)$ | $u_2(4)$ | $u_3(6)$ | No Choice |
| 4 | $u_1(4)$ | $u_2(5)$ | $u_3(7)$ | No Choice |
| 5 | $u_1(5)$ | $u_2(6)$ | $u_3(1)$ | No Choice |
| 6 | $u_1(6)$ | $u_2(7)$ | $u_3(2)$ | No Choice |
| 7 | $u_1(7)$ | $u_2(1)$ | $u_3(3)$ | No Choice |

*Identify any constraints on the mixture components or other factors in order to specify the experimental region and the response variable(s) to be measured*

As can be seen in Table 8.1, the mixtures show the proportions of the budget allocation to the attributes. In each mixture, the sum of attributes should be equal to the budget constraint. Furthermore, amounts to be allocated should be restricted in order to not dominate other choice alternatives. Therefore, we apply $x_{max} * u_{max} \leq u_{min}$ where $u_{max} = max\ (u_i)$ and $u_{min} = min\ (u_i)$. This restriction depends on the mixture level used in the experiment as $x_{max}$ is the maximum mixture level and $x_{max} < 1$ (Raghavarao and Wiley 2009). As we consider three-attribute mixtures, our restriction will be $2/3 * u_{max} \leq u_{min}$.

*Propose an appropriate model for modelling the response data as functions of the mixture components and other factors selected for the experiment*

We use a second degree mixture model for a given amount, which allows interaction between amounts and mixture proportions. This model allows us to determine the optimum amount in the range of experimentation. Equation 4 shows the second degree mixture-amount model.

$$E(y) = \alpha u + \beta u^2 + \sum_{i=1}^{q-1} \gamma_i u x_i + \sum_{i=1}^{q} \delta_i x_i + \sum_{\substack{i,j=1 \\ i<j}}^{q} \delta_{ij} x_i x_j \qquad (4)$$

where *E(y)* is the rating response or choice response of a mixture $(x_1, x_2 \ldots x_q)$ with given amount of u. In addition, each attribute has its own attached levels, bringing additional utility. Effect coding is used to measure the non-linear effects of attribute levels. Equation 5 shows the utility of levels in the case of three three-level attributes.

$$E(t) = \beta_0 + \beta_{11}t_{11} + \beta_{12}t_{12} + \beta_{21}t_{21} + \beta_{22}t_{22} + \beta_{31}t_{31} + \beta_{32}t_{32} \quad (5)$$

where $t_{11}$ and $t_{12}$ show the effect of the first and second saving time levels for the first attribute, $t_{21}$ and $t_{22}$ are related to the first and second saving time levels of second attribute and $t_{31}$ and $t_{32}$ have the same definition for the third attribute. Accordingly, our utility function is equal to:

$$U_T = E(y) + E(t) \quad (6)$$

*Select an experimental design that is sufficient not only to fit the proposed model, but which allows a test of model adequacy as well*

To examine how these allocations may be influenced by corresponding benefits (captured in terms of attributes), the three attribute levels for each of the three pricing options should also be varied. In stated preference experiments, orthogonal fractional factorial designs are commonly constructed. However, the issue of orthogonality is problematic in the current situation. Since the mixture amount needs to meet the budget constraints, these designs are in general non-orthogonal. Likewise, due to the number of choice sets, it is impossible to create an experimental design for the attribute levels that is both orthogonal and in relation to the mixture amount design. In the present study, it has therefore been decided to create a Latin Square design for 3-level factors.

## Example

In order to understand travellers' response to accumulated pricing policies, we designed a mixture-amount experiment to test our design. In this experiment, governments stimulate people to use transport modes other than cars to go to work, study, and so on. For this reason, they want to implement three kinds of pricing policies on traveller's routine routes from home to their destination: toll roads, congestion pricing and parking policies. If people indeed choose one or more of these options, they will benefit from these policies by saving travel time and arriving at work or home earlier. Their savings in travel time may depend on what they pay for and how much they pay though. So, governments can benefit from understanding how many people would be willing to pay for different gains in travel time and how they might allocate their budget to the three options which may differ in costs and/or travel time savings.

Since we are interested in how sensitive travellers are to price level and whether the same preferred mixture of allocations applies at different expenditure levels, we consider three levels of amount in a day ($u_1$, $u_2$, and $u_3$) that users can spend on these policies in daily travel: €6, €7.50, and €9. We also consider three time levels: 5, 10 and 15 minutes, which users can save every day during trips in accordance with their willingness to pay for accumulated pricing policies. As mentioned earlier, we attached these levels to the attributes using a Latin square design.

**Table 8.3 Mixture profiles for the three budget amounts**

| Profile | Road Toll ($x_1$) | | Congestion Pricing ($x_2$) | | Parking Tariff ($x_3$) | | Total | | Rating |
|---|---|---|---|---|---|---|---|---|---|
| | Travel Cost (€/day) | Saving Time (min) | Travel Cost (€/day) | Saving Time (min) | Travel Cost (€/day) | Saving Time (min) | Travel Cost (€/day) | Saving Time (min) | |
| 1 | 6 | 5 | – | – | – | – | 6 | 5 | 9 |
| 2 | – | – | 6 | 15 | – | – | 6 | 15 | 7 |
| 3 | – | – | – | – | 6 | 15 | 6 | 15 | 10 |
| 4 | 4 | 15 | 2 | 15 | – | – | 6 | 30 | 8 |
| 5 | – | – | 4 | 10 | 2 | 5 | 6 | 15 | 8 |
| 6 | 2 | 5 | – | – | 4 | 10 | 6 | 15 | 9 |
| 7 | 2 | 15 | 2 | 10 | 2 | 15 | 6 | 40 | 8 |
| 8 | 7.50 | 5 | – | – | – | – | 7.50 | 5 | 7 |
| 9 | – | – | 7.50 | 5 | – | – | 7.50 | 5 | 4 |
| 10 | – | – | – | – | 7.50 | 5 | 7.50 | 5 | 8 |
| 11 | 5 | 10 | 2.50 | 15 | – | – | 7.50 | 25 | 6 |
| 12 | – | – | 5 | 5 | 2.50 | 15 | 7.50 | 20 | 5 |
| 13 | 2.50 | 15 | – | – | 5 | 5 | 7.50 | 20 | 8 |
| 14 | 2.50 | 10 | 2.50 | 10 | 2.50 | 5 | 7.50 | 25 | 6 |
| 15 | 9 | 5 | – | – | – | – | 9 | 5 | 4 |
| 16 | – | – | 9 | 10 | – | – | 9 | 10 | 1 |
| 17 | – | – | – | – | 9 | 15 | 9 | 15 | 6 |
| 18 | 6 | 15 | 3 | 5 | – | – | 9 | 20 | 3 |
| 19 | – | – | 6 | 5 | 3 | 5 | 9 | 10 | 2 |
| 20 | 3 | 10 | – | – | 6 | 10 | 9 | 20 | 5 |
| 21 | 3 | 10 | 3 | 5 | 3 | 15 | 9 | 30 | 4 |

To show the approach, parameter values were assumed. Based on these assumed values, artificial data were created (Table 8.3) using the mixture values from Table 8.1. Accordingly, seven choice sets given in Table 8.2 were rated on a 1–10 scale ranging from 1 (poor) to 10 (excellent). Ties were also allowed and coefficients were estimated based on this artificial rating data provided in Table 8.3. Table 8.3 shows the mixture profiles for the three budget amounts as well as their associated travel time savings.

The first seven profiles involve the amount of €6. Profiles 8–14 and 15–21 relate to the amounts of €7.50 and €9 respectively. Regression analysis was applied to show that these input data can be recovered. Because standard estimates and t-values don't have much meaning in this context, we have only reported estimated coefficients. Equation 7 shows the final result of OLS.

$$\begin{aligned} U_T = &+0.9u - 0.15u^2 - o.15ux_1 - 0.45ux_2 - 0.3x_1 - 0.8x_2 - 0.15x_3 \\ &-0.88x_1x_2 - 0.05x_1x_3 + 0.01x_2x_3 \\ &+10 - 0.05t_{11} + 0.02t_{12} - 0.08t_{21} + 0.01t_{22} - 0.06t_{31} + 0.02t_{32} \end{aligned} \tag{7}$$

As Equation 7 shows, the first-order term of the travel budget has a positive sign, while the quadratic term has a negative sign. This is the expected form for the second-degree canonical model and suggests a decreasing utility with increasing amounts. The negative effects of toll roads, congestion pricing and parking price in the model show that utility decreases if the share in the mixture for each of these policies increases. It also shows that this reduction is higher for congestion pricing compared to other policies.

In the context of saving time as a result of the various pricing policies, the results indicate that the first saving time level (5 min) has the lower utility and second and third levels (10 and 15 min) have the higher utility compared to the mean. As expected, the highest saving time level has the highest utility compared to the other levels. Figure 3 shows the relation between total utility ($U_T$) and travel budget for each profile.

Figure 8.3 shows that, as expected, utility decreases with increases of the travel budget spent on pricing schemes. In general, the second profile reflecting congestion pricing seems to have the lowest utility while the third profile which shows the parking has the highest utility compared to the other profiles. Figure 8.4 also demonstrates the relation between total utility and profiles for each travel budget level.

As Figure 8.4 shows, similar patterns can be seen for different budget levels regarding different profiles. Also, as expected, the highest and lowest budget levels have the lowest and highest utility respectively. Between combined profiles (4, 5, 6 and 7), which reflect the combination of three mentioned policies, profile 6 (which is related to the combination of toll roads and parking) has the highest utility for all travel budget levels.

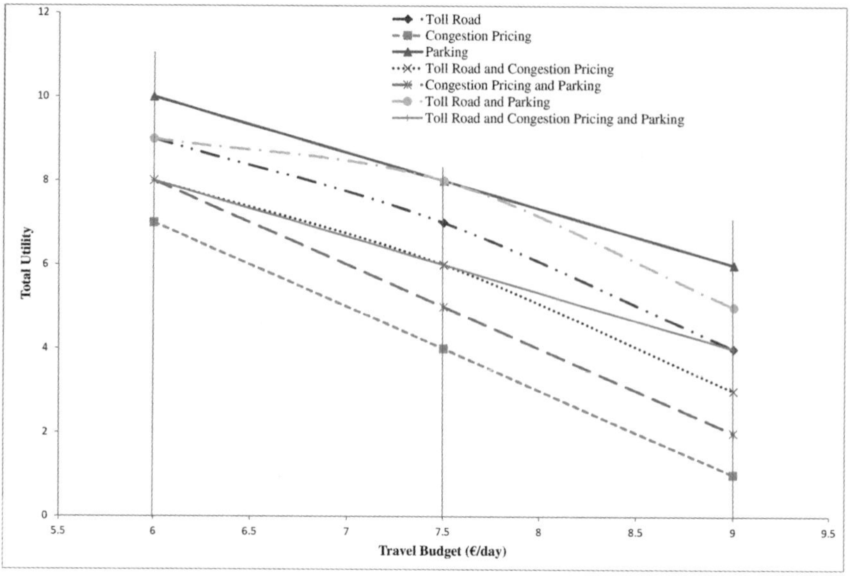


**Figure 8.3 Relation between $U_T$ and budget levels for each profile**

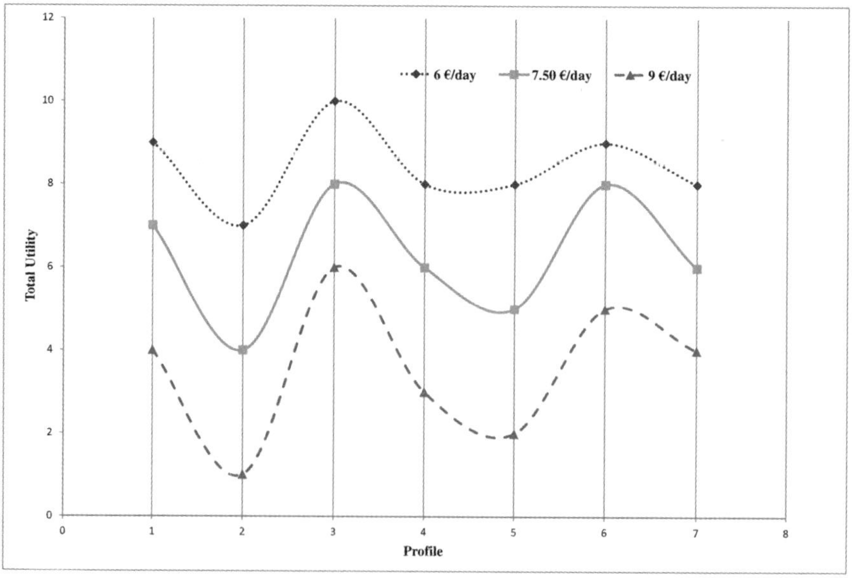


**Figure 8.4 Relation between $U_T$ and profiles for each budget level**

## Summary and conclusions

Reducing congestion, delays and emissions as well as increasing revenue in urban areas contributes to sustainable mobility. These are oftentimes the main objectives of pricing policies. This chapter attempted to assess the effects of accumulated pricing policies on travel behaviour and therefore sustainable mobility. To address this issue, we have described an experimental approach to better understanding how travellers might respond to such accumulated pricing schemes, subject to budget constraints. This study applied a mixture-amount design in the case of pricing policies in order to look at the effect of changing both the amount and the mixture of attributes. To the best of our knowledge, it is one of the first applications of mixture-amount choice experiments in transport analysis. Moreover, by considering the effect of varying attributes levels, we extended the classical mixture-amount design. In this, a mixture-amount experiment was assumed, with 3 attributes (toll road, congestion pricing and parking price) and three different daily travel budget levels. Additionally, each pricing policy had three levels of saving travel time. Seven mixtures of the simplex lattice design {3, 3} were used for the allocation process and a second-degree polynomial model was used for the modelling part. Finally, the methodology was tested with an example using artificial data. The results showed the potential of this proposed approach in terms of road pricing studies. It revealed a negative tendency to pay for pricing policies and showed that in all travel budget levels of the study, parking has the highest utility and congestion pricing has the lowest.

In future analyses, it may be relevant to test the approach with real data in disaggregate levels and to use choice experiment in order to capture the effects of every parameter. This can be done more precisely by using advanced discrete choice models like Mixed Logit (ML). Additionally, socio-economic variables should be taken into account in the next step of analysis, in order to test how the results are affected by respondent characteristics.

## Acknowledgements

The research leading to these results has received funding from the European Research Council under the European Community's Seventh Framework Programme (FP7/2007–2013) / ERC grant agreement n° 230517 (U4IA project). The views and opinions expressed in this publication represent those of the authors only. The ERC and European Community are not liable for any use that may be made of the information in this publication.

**References**

Arentze, T., Hofman, F. and Timmermans, H. 2004. Predicting multi-faceted activity-travel adjustment strategies in response to possible congestion pricing scenarios using an Internet-based stated adaptation experiment. *Transport Policy*, 11, 31–41.

Bhat, C.R. and Castelar, S. 2002. A unified mixed logit framework for modeling revealed and stated preferences: formulation and application to congestion pricing analysis in the San Francisco Bay area. *Transportation Research*, 36B (7), 593–616.

Burris, M.W. and Pendyala, R.M. 2002. Discrete choice models of traveller participation in differential time of day pricing programs. *Transport Policy*, 9 (3), 241–51.

Colbourn, C.J. and Dinitz, J.H. 2007. *Handbook of Combinatorial Designs*. 2nd edition. Boca Raton: Chapman & Hall / CRC.

Danwen, B., Wei, D. and Shihui, G. 2010. Impact of parking rates on resident travel behaviour. *Journal of Transportation Systems Engineering and Information Technology*, 10 (3), 80–85.

Dissanayake, D. and Kouli, S. 2007. *Intercity Route Choice Stated Preference Model to Investigate Driver Response to Road Pricing: Proceedings of the 86th Annual Meeting of the Transportation Research Board*. Washington, D.C., January 2007.

Guan, H., Sun, X., Liu, X. and Liu, L. 2005. *Modeling Parking Behaviour for Better Control and Pricing: A Case Study from One of the Busiest Retail Shopping Areas in Beijing, China: Proceedings of the 84th Annual Meeting of the Transportation Research Board*. Washington, D.C., January 2005.

Keuleers, B., Chow, V., Thorpe, N., Timmermans, H. and Wets, G. 2005. Behavioural change in activity-travel patterns in response to road user charging. *Journal of Transport Economics and Policy*, 40 (1), 119–34.

Peters, J.R., Shim, H.S. and Kress, M.E. 2011. *Disaggregate Multimodal Travel Demand Modeling Based on Road Pricing and Access to Transit: Proceedings of the 90th Annual Meeting of the Transportation Research Board*. Washington, D.C., January 2011.

Piepel, G.F. and Cornell, J.A. 1994. Mixture experiment approaches: examples, discussion, and recommendations. *J. Quality Technology*, 26, 177–96.

Raghavarao, D. and Wiley, J.B. 2009. Conjoint measurement with constraints on attribute level: a mixture-amount model approach. *International Statistical Review*, 77 (2), 167–78.

Raghavarao, D., Wiley, J.B. and Chitturi, P. 2011. *Choice-Based Conjoint Analysis: Models and Designs*. USA: Taylor & Francis Group, LLC.

Scheffe, H. 1958. Experiments with mixtures. *Journal of the Royal Statistical Society, Series B (Methodological)*, 20 (2), 344–60.

Stinson, D.R. 2003. *Combinatorial Designs: Constructions and Analysis*. New York: Springer.

Yan, J., Small, K. and Sullivan, E. 2002. *Choice Models of Route, Occupancy, and Time-of Day with Value Priced Tolls*: *Proceedings of the 81st Annual Meeting of the Transportation Research Board*. Washington, D.C., January 2002.

# Chapter 9
# Mobility Biographies: Studying Transport and Travel Behaviour Through Life Histories

Andrew Miles, Niamh Moore and Stewart Muir

## Introduction

As the negative environmental effects of carbon-fuelled hypermobility on the environment become increasingly manifest, there is a pressing need to better understand how significant changes in personal mobility might be brought about. Such knowledge might then be utilized by transport planners to develop models more responsive to the dynamics of people's lives. Many model-based desk studies have aimed to simulate sustainable urban environments as part of a decision support tool. However, they have adopted a number of unvalidated, hypothetical assumptions in the process, particularly in the way major transport-focused interventions might impact behaviour and the effectiveness of infrastructure. As a result, the ability to apprehend and predict change is limited (Flyvbjerg et al. 2006).

In this chapter, we assess the promise of the 'mobility biographies' approach for providing an alternative way of understanding the dynamics of mobility practices in contrast to those routinely adopted by transport modellers. While welcoming the emergence of this approach within transport studies, we argue that it has yet to reach fruition because the full implications of this approach remain underdeveloped. We suggest that this limitation of potential relates to the challenges of exploiting specifically qualitative research approaches in the context of a domain, transport studies, which has largely been dominated by quantitative research paradigms and thus remains distant from a rich tradition of biographical and narrative research. Our intervention here emerges from the early stages of a transdisciplinary research project that uses a pioneering longitudinal and qualitative household panel study on travel and transport behaviour to help develop more sensitive modelling tools to effectively plan practices. Part of this project involves reworking the theory and method of mobility biographies.

## The limitations of contemporary transport modelling

Since the 1950s, the development of academic transport demand modelling focused on predicting long-term change has been influenced by two broad theoretical narratives; both working with fixed representations of individual/ist behaviour

(Pas 1990, Timms 2008). The first, which was particularly influential up until the 1970s, draws on analogies from Newtonian physics to describe the properties of transport systems and applies the metaphor of electrons within electrostatic fields to human actors and their environments. This narrative was subsequently supplanted by approaches rooted in neo-classical economics, representing transport users as rational-choice directed, utility-maximizing individuals operating within a state of equilibrium and which largely continue to hold sway. Meanwhile, at the other end of the spectrum to the academic modelling of long-term change, and partially in response to its limitations, the mathematical models used by practical transport studies for city transport planning continue to focus on stable situations and slow change, with existing transport surveys probing marginal trade-offs rather than major qualitative change.

More recently, several approaches have sought to problematize and challenge the models of transport behaviour drawn from economics. For example, the debate surrounding the value of travel time savings has explored the psychological components of transport choices and decision-making (Lyons and Urry 2005). This includes work by Mokhtarian et al. (2001) on the positive utility of travel and by Jains and Lyons (2007) on travel as a dimension of social reciprocation in networks to illustrate the complexity of relationships between travel experience and time valuations. There is also a growing body of research on the social impacts and distributional effects of travel behaviour (Parkhust and Shergold 2009, Markovich and Lucas 2001), including the putative role of subjective well-being in transport policy (Delbosc 2012). While these are welcome interventions, none of these studies have attempted to grasp the complex temporalities of people's lives. As a result, they cannot address issues of how and why significant change in people's travel behaviours might occur.

Clifton and Handy have noted that transportation research 'has been entrenched in a quantitative paradigm for some time' and have linked this to the historical dominance of science and engineering in shaping the field (2001: 16). We would also link this to the emphasis on economic models. In reviewing developments in qualitative research on travel behaviour, Clifton and Handy have gone so far as to insist that 'without more widespread use of qualitative techniques in travel behaviour research, we will make little meaningful progress towards improving our fundamental understanding of travel behaviour' (2001: 17). Their cautionary observation that 'doing qualitative research is more challenging than transportation researchers might think' informs our efforts to best exploit the potentials of mobility biographies.

While specific literature on mobility biographies is the main frame for this chapter, we also see in our approach the basis of a broader, critical but constructive, response to calls for a new 'mobile sociology' (Urry 2000, 2007). Firstly, in the sense that mobility biographies provide a novel and substantial method for not only representing but demonstrating and making sense of the 'interdependence' of multiple mobilities (Urry 2000, Larsen et al. 2012). Secondly, as a corrective to existing accounts within the mobilities field, which until now have failed to embrace and develop a more complex account of temporal (as well as spatial) change.

## Mobility biographies

One promising approach to understanding the process of behaviour change in the context of transport and travel practices is the nascent 'mobility biographies' paradigm in transport studies. This term has been used by several authors in this field, including Axhausen (2008), Frändberg (2008), Scheiner (2007) and especially Lanzendorf (2003, 2010).

Focusing on the inability of current approaches to address dynamic change in people's lives, Lanzendorf (2003) identifies three limitations in the existing research: (i) the dearth of research on the effects of long-term decisions on travel behaviour; (ii) the reliance of much travel research on cross-sectional rather than longitudinal data; and (iii) a focus on statistical correlations between factors rather than on causal relationships. Lanzendorf identifies three methods which would address these shortcomings: pseudo-panels from time series data, panel data and what he calls 'qualitative retrospective' data. In regards to the first method, he argues that pseudo-panels rely too much on aggregating data and are therefore not useful for understanding individual behaviour change. Although more positive about the possibilities offered by panel data, he feels that issues such as panel attrition and the timescale and costs of the approach are significant. Finally, he concludes that while similar challenges are posed by qualitative retrospective methods, they are considerably less. It is from this analysis of the field that Lanzendorf then turns to the possibilities of mobility biographies.

This attempt to overcome the problematic framing of conventional transport research is a welcome intervention. However, this contribution and the body of work that has developed around mobility biographies have arguably, so far, fallen short of exploiting the potential of a more thoroughly qualitative approach. We note the persistence of a positivist approach, evident in Lanzendorf's (2003: 2) original conception of a mobility biography as:

> the total of an individual's longitudinal trajectories in the mobility domain [including the assumption] that events in these trajectories exist, or, put in other words, that at certain moments in individual's life the daily travel patterns, the car ownership or other mobility characteristics change to an important degree.

While drawing on a life course approach offers some new perspectives with respect to the dynamics of travel behaviour, we suggest that the possibilities of mobility biographies extend beyond an understanding of people's behaviour as 'subdivided into a series of life course trajectories: events and stages between events in certain domains of life', with the principal interest of life course researchers being 'to analyse the probability of interactions in different life course trajectories' (Lanzendorf 2003: 9).

In this account of mobility biographies, the different affordances of an explicitly narrative, life history approach to mobility biographies, as distinct from a life course approach, are not explored. Adhering to the dominant methodological framework in transport studies, it neglects extensive engagements with narrative and life stories undertaken in the broader social sciences and humanities (Riessman 1993, Andrews et al. 2000, Riessman 2002, Roberts 2002, Chamberlayne et al. 2000). This absence is evident in the way that existing mobility biographies research tends to adopt a realist approach to biographical accounts, extracting content from interviews in ways which (i) demonstrate a linear approach to people's lives, (ii) divide people's lives into events and stages and (iii) conceptualize causality as a linear process in which an event in the past (such as having children) leads to a change in behaviour (buying a car) that can be recalled and recounted in the present. As a consequence, research in this area often makes considerable effort to address concerns about the reliability, and therefore validity, of memory. In contrast, for many interpretative researchers the value of a biographical approach lies precisely in its access to subjective accounts of the past as a resource for better appreciating the ways in which people make meaning of (and justify) their lives.[1]

## The Step-Change project

Our assessment of the promises and limitations of the mobility biographies approach to understand and model change emerges from our work on the Step-Change project.[2] This five-year project is an ambitious interdisciplinary collaboration of transport engineers, mathematicians and social scientists and is designed to examine how we can bring about a 'step change' in the way we understand change in travel behaviour. This also includes an understanding of the way these changes are represented in transport models and thus translated into planning practices. The research programme of this project is structured around five inter-related strands of work (Figure 9.1).

1 Moreover, both the framing of existing work in terms of individual biographies, and the ways in which they are graphically represented and described (see, for example, Frändberg 2008), tends to underplay the recursive relations between people's actions, social relationships and the physical environment.

2 'Sustainable Transport Evidence and modelling Paradigms: Cohort Household Analysis to support New Goals in Engineering design' (STEP-CHANGE). Funded by the Engineering and Physical Sciences Research Council (EP/I00212X/1). Principal Investigator M. Tight (Birmingham). Co-Investigators: A. Miles, N. Moore (Manchester), M. Savage (LSE), P. Timms, D. Watling (Leeds). See www.changing-mobilities.org.uk. We would like to thank our collaborators on the STEP-CHANGE project for their comments on earlier drafts of this chapter.

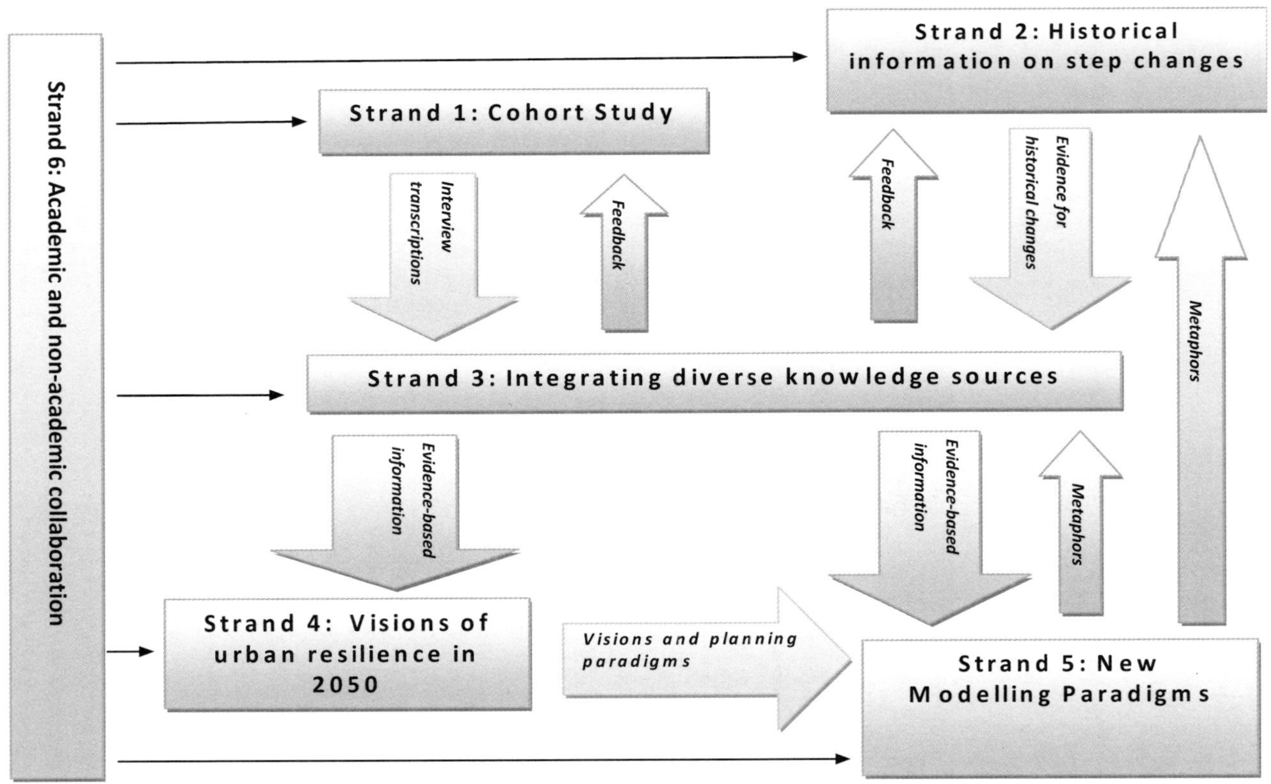


Figure 9.1 The Step-Change project work map (figure courtesy of Paul Timms)

Our arguments in this chapter specifically draw on data emerging from Strand 1, from the first wave of a qualitative longitudinal panel study focusing on travel and transport issues, which provides the empirical base-frame of the Step-Change project. The panel is comprised of 240 participants from households in contrasting urban areas in Leeds and Manchester, two post-industrial cities in the northern United Kingdom. Four recruitment locations in each city were chosen using a sampling matrix which combines high, middle and lower income areas, types of transport infrastructure, proximity to city centres (such as central, suburban and peri-urban zones) and relative demographic stability versus changeability. This strategy allows for the systematic comparison of residents from similar kinds of places in two different urban environments and allows for the exploration of the effects of the physical transport context.[3]

The panel was recruited via a combination of telephone contact, door-knocking and snowballing sampling methods with participants chosen on the basis of their residence in the selected study areas. A sample of 30 participants was chosen from each of the eight locations to participate in annual qualitative, in-depth, semi-structured interviews and to independently complete a seven-day diary of all movement outside the house.[4] In this way, the cross-sectional travel data collected in the travel diaries can be analysed longitudinally and changes in travel rhythms over the course of the study can be explored in the qualitative interviews. An effort was made to secure a sample that approximately matched the age and ethnicity profile of each study area whilst allowing for the recruitment of demographic outliers. Although 240 participants is a large sample for a qualitative study (see Mason 2010) it is manifestly small in comparison to major quantitative studies. As such, the sample is not intended to be statistically representative of the general population or generalizable for different 'types' of people based on specific variables or characteristics. Rather, as is common in qualitative research, the sampling strategy allowed for an in-depth exploration of a range of different travel and life experiences as well as an exploration of the effects of a range of social and environmental processes across and within the eight geographical areas (Mason 2002: 120, 123–7).

In essence, the purpose of the panel study is to use the cohort to situate changes (and continuities) in people's mobilities (and immobilities) in the context of their specific location, generational position, background, life circumstance, social

3 The type and number of study areas were also chosen to allow for comparison with previous qualitative studies of local areas such as Savage et al.'s (2005) research on experiences of socio-economic and cultural change across Manchester (also see Elliot et al. 2010).

4 In future waves it is planned that a sub-sample of panel members will also take part in ethnographic encounters. This research phase is likely to involve, amongst other things, fieldwork in participant neighbourhoods and may include project data collected 'on the move' as researchers shadow selected participant journeys in a way that supplements the supposedly 'static' process of interviewing (see Büscher and Urry 2009: 103).

relationship, life history and day-to-day life. It does so by ambitiously combining two of the promising approaches identified by Lanzendorf (2003): a household study undertaken over several years and a substantive focus on mobility over the life course. The longitudinal element of the study, although covering a relatively short time period of four years, allows for the observation of short to medium term patterns of change (or lack of) over the course of the project. These are dynamics that can be explored in the qualitative interviews as the study progresses.[5]

The first wave of interviews included an in-depth discussion of each participant's life history, memories of travel and transport throughout their life, current travel routines and the emotional and symbolic meanings of particular journeys and modes of travel. They also explored the participants' broader histories of 'movement and settlement' (Daniels and Nash 2004: 449). Beginning with an account of residential movement and 'belonging' through their life, participants were then asked to recount their 'life story'. This was set up as an open invitation, so that participants themselves determined the content and shape of their accounts. It was important that each interview was not limited to a discussion of transport activities. Instead, the interviews sought to locate transport activities in the context of members' life histories, associations and everyday practices. Only once they had recounted their stories were respondents then asked to specifically identify and discuss key life stages or events and their relationship to travel and transport. This open-ended discussion structure, followed by more specific questioning on key moments or phases, crucially allowed the interview to move beyond a focus on conventional life-cycle 'stages' (such as leaving school or retiring) favoured by previous attempts at mobility biographies. This provides for an exploration of the times, events and occurrences that were important to participants (for example, joining a dance class or getting a dog) as well as how a participant's life may be recalled, recounted and accounted for, in ways that do not always accord with a linear series of causal 'events'.

Step-Change methodology is also innovative in that it explicitly aims to elicit and generate multiple versions of people's biographies. Much interpretive qualitative research draws on the assumption that a person's account of their life is (co-) produced through the relationship between the speaker and the listener. Thus, accounts can change from one telling to the next, and in relation to the different ways in which questions are asked (Mason 2002, Byrne 2003). It is less common though to attempt to explore multiple ways of constructing a life in a single interview. Also, it is worth noting here that while travel and mobility are widely assumed to be embedded in everyday life, they are rarely considered in these terms. This reflects, in part, the difficulty of eliciting mobility histories as part of more general biography, which in turn is perhaps because so much about mobility is felt to be routine, mundane and therefore taken for granted (Aldred and Jungnickel 2012: 523–4). The aim of the Step-Change approach is not to

---

5 The data from the study will also be archived and available for reuse through www.timescapes.leeds.ac.uk.

triangulate these accounts into a closer approximation of a factually accurate history, but rather to gain a sense of the way in which perceptions of the past, and of life as an ongoing process, are drawn upon in participants' experiences and accounts of the present (Freeman 2010).

## Re-configuring mobility biographies

A key innovation of the Step-Change project is our development of existing work in transport studies on 'mobility biographies'. Here, we bring this approach into further conversation with an extensive history of qualitative research involving biography, personal narrative and life history, as well as a specific emerging literature around sustainability, biography and everyday life (for example, Pink 2012, Pooley et al. 2005).[6] This section highlights some of the key issues concerning realism, linearity, the problems related to a focus on events and transitions, and a search for causality, that an interpretive qualitative research perspective raises about the mobility biographies approach, as it is presently framed within the transport studies discipline. We use a case study from one of our respondents interviewed during the first wave of the Step-Change panel study.

Specifically, we illustrate and develop the points above by referring to the example of 'Laura', a 40-year-old, white British woman living in north Manchester. Clearly her account is not intended to be representative or generalizable, nor is it chosen to offer a definitive account of mobility or mobility biographies. Rather, we use it as an examplar to show the limits of the current mobility biography approach, as well as the possibilities of a more interpretative narrative approach and as a demonstrator of our method. We chose this case study because Laura is in part atypical, meaning that she does not own a car or drive one, rather she has a strongly articulated identity as a walker. At the same time, as described below, her account undoes any straightforward reading of her identity as a walker. Furthermore, in seeking to provide a more complex account of temporality in people's narratives, Laura offers an exemplary illustration of the non-linearity of time in some people's lives and the alternative ways they experience and recount their experiences. While Laura's refusal of linear time is particularly clear for illustrative purposes, it is not inconsistent with other narratives. Laura may be atypical in terms of her lack of daily car use, but her narrative strategies are not. Focusing on one respondent also enables us to explore the complexities of one narrative as well as the affordances of the method. This would not be possible by drawing on shorter quotes from a series of interviews (Thomson 2007, Taylor 2012).

6 See also the ESRC/EPSRC funded 'Energy Biographies: Understanding the Dynamics of Energy Use for Energy Demand Reduction', project based at Cardiff University http://www.energybiographies.org.

One of the striking things about Laura's interview is the way she very clearly, often emphatically, identified herself at numerous points as a walker. This was repeated throughout the interview:

> 'But I walk nearly everywhere …'

> 'But for me, you know, as long as like my legs are walkable, I will walk.'

> 'But I'm one who walks [laughs] I've always been a keep fit fanatic as well so I try and look after myself by walking everywhere I go.'

> 'Always been a walker. No, never had me own car, no.'

> 'Yeah, I mean I can't afford a car, it's too expensive. I'd sooner –, for what it costs to keep a car you might as well catch a bus or a Metro, you know, it's just as [inaudible] but the majority of the time I do walk wherever I go.'

Laura's account of her travel was also supported by the cross-sectional data emerging from her week-long travel diary. This showed her relatively limited movement, almost all of it local, and her reliance on walking as a primary means of transport (see Figure 9.2).

Further, when selecting the diagram she felt best represented her life as a whole (see Figure 9.3), Laura also chose to emphasize a lack of significant movement, of lasting change, in her life and to suggest her continual movement back to the mean.

To take a realist approach to Laura's account, or in other words to accept the way she presents herself at face value – that she is a walker, and has always been a walker and that there are no transitions to, or through, other types of transport use – would not do justice to the complexity of her narrative. Initially, when asked to recount her life story, she repeatedly stressed not only the fallibility of her memory, but also that it was completely limited:

> 'I can't remember from being a kid, there's no way I can talk about being a kid. I can't remember it.'

> 'I say that to my mum, I can't remember from being four, I can't remember from being ten, I can't remember from being anything of being a kid. I can't.'

> 'You know, I can't even remember my dad, you see my dad's not with us anymore, my dad died 18 months ago.'

STEP-CHANGE PROJECT: TRAVEL DIARY    Day of week THURSDAY    Date 15/12/2011

| Journey | Journey | Journey | Journey | Journey | Journey | Journey |
|---|---|---|---|---|---|---|
| **Mode used** (please number in order of use)<br>Car<br>Bus<br>Train<br>Cycle<br>Walk 1<br>Motorcycle<br>Other: | **Mode used** please number in order of use)<br>Car<br>Bus<br>Train<br>Cycle<br>Walk 1<br>Motorcycle<br>Other: | **Mode used** (please number in order of use)<br>Car<br>Bus<br>Train<br>Cycle<br>Walk 1<br>Motorcycle<br>Other: | **Mode used** (please number in order of use)<br>Car<br>Bus<br>Train<br>Cycle<br>Walk<br>Motorcycle<br>Other: | **Mode used** (please number in order of use)<br>Car<br>Bus<br>Train<br>Cycle<br>Walk<br>Motorcycle<br>Other: | **Mode used** (please number in order of use)<br>Car<br>Bus<br>Train<br>Cycle<br>Walk<br>Motorcycle<br>Other: | **Mode used** (please number in order of use)<br>Car<br>Bus<br>Train<br>Cycle<br>Walk<br>Motorcycle<br>Other: |
| **Origin**<br>HOME<br>**Destination**<br>CAFE<br>**Purpose**<br>LUNCH. | **Origin**<br>CAFE<br>**Destination**<br>TESCO.<br>**Purpose**<br>SHOPPING. | **Origin**<br>TESCO.<br>**Destination**<br>HOME<br>**Purpose**<br>GET HOME | **Origin**<br>**Destination**<br>**Purpose** | **Origin**<br>**Destination**<br>**Purpose** | **Origin**<br>**Destination**<br>**Purpose** | **Origin**<br>**Destination**<br>**Purpose** |
| **Journey time**<br>Depart: 1-00 ~~am~~/pm<br>Arrive: 1-05 ~~am~~/pm | **Journey time**<br>Depart: 4-00 ~~am~~/pm<br>Arrive: 4-10 ~~am~~/pm | **Journey time**<br>Depart: 5-00 ~~am~~/pm<br>Arrive: 5-15 ~~am~~/pm | **Journey time**<br>Depart: ______ am/pm<br>Arrive: ______ am/pm | **Journey time**<br>Depart: ______ am/pm<br>Arrive: ______ am/pm | **Journey time**<br>Depart: ______ am/pm<br>Arrive: ______ am/pm | **Journey time**<br>Depart: ______ am/pm<br>Arrive: ______ am/pm |
| **Distance** (~~in miles~~)<br>300 M. | **Distance** (~~in miles~~)<br>300 M. | **Distance** (~~in miles~~)<br>300 M | **Distance (in miles)** | **Distance (in miles)** | **Distance (in miles)** | **Distance (in miles)** |
| **Travelling with:**<br>Alone 1<br>Partner<br>Family<br>Friends<br>Others: | **Travelling with:**<br>Alone<br>Partner<br>Family<br>Friends 1<br>Others: | **Travelling with:**<br>Alone 1<br>Partner<br>Family<br>Friends<br>Others: | **Travelling with:**<br>Alone<br>Partner<br>Family<br>Friends<br>Others: | **Travelling with:**<br>Alone<br>Partner<br>Family<br>Friends<br>Others: | **Travelling with:**<br>Alone<br>Partner<br>Family<br>Friends<br>Others: | **Travelling with:**<br>Alone<br>Partner<br>Family<br>Friends<br>Others: |

**Figure 9.2** Laura's travel diary

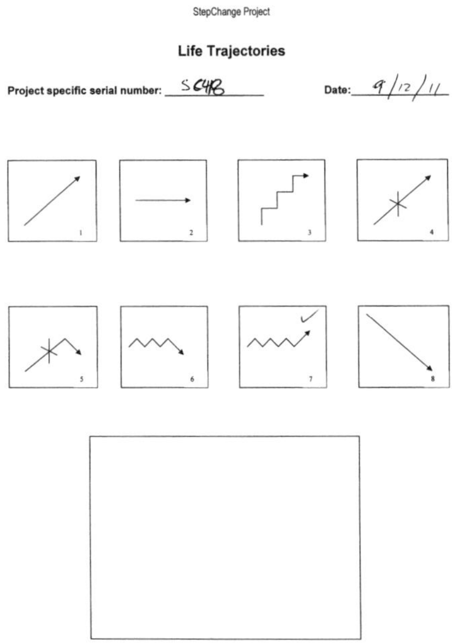


'I don't know because it seems like an even keel, you know, up, down, up, down. 'Cause everyone has little ups and downs, ups and downs, but then the majority of the time I always get back on my feet. I always end up on my feet. So I'd say the up would be –, however up and down, how down ever I go I always end up going back up again, down again … and I always end up on my feet and always end up on top of cloud nine. 'Cause only you can make your life work, no one else can.'

**Figure 9.3 Laura's chosen Life Trajectories diagram**

Were we to have taken her own account literally we might, on Laura's own admission, discount her entire interview. This is the implication of Lanzendorf's and others' concerns about the reliability, and hence validity, of memory. Yet, bearing in mind that *how* a person's life is recounted is precisely a resource for many qualitative researchers, Laura's insistence on the fallibility of her memory is rather, in itself, to be understood as important data. Furthermore, despite the repeated insistence on her own lack of memory and her own admission of unreliability, Laura was able to provide an extensive account of her school and work histories when prompted:

> 'I went to school, I went to college, did a diploma for four years in computer studies from 2001 to 2005 here at Abraham Moss on Cheetham village. I did a course there for four years. Went to college and then after that did an administration NVQ admin course for about a year, [laughs] it's not things I think about, I don't think of the past, I think of the future. I don't live in the past. […] I went to school, I left school, couldn't get a job at 17, and if I can remember. You go in the Jobcentre, there's nothing there, they're asking for people who are older, 20-odd, and then when I got to 20-odd they were asking for people at 17 because they were asking for less pay. Right. So that's one thing. Then I'm looking for a job, I used to work in a butcher's, Saturday jobs, things like that. Erm, trying to think. Left there, just generally couldn't get a job. I went back to college, did a typing course 'cause I didn't do typing at school.

> Went back to college, did a typing course, learnt to type, NVQ in typing I think or diplomas, stuff like that. If I can remember that must have been in my 20 –, well it must have been 'cause I left school at 16. It must have been 18 to 20. I went back to college. I've always been in college, I've always done something to teach meself something. That was about 20. I went to get a job in 19 –, how old was I? 1991, erm, never been in work very much because work's been really hard to get. Saturday jobs or butchery, I used to work in butcher's and markets in [inaudible 17:25]. And eventually I got a job in Birthdays in Manchester. So that was where I worked for seven years. After that I left there in 1998 with depression, never went back to work, I left after a year. From there I went back to college. So I did my –, that's when I decided I wanted –, computers were coming in and typing wasn't in so much. So I went back to college for four years and I did a diploma then. And then from there I worked for –, I got a job eventually with DWP in 2003. After I left or it must have been while I was working, doing still the course. And I was there for year. Got sciatica. I left the job because they didn't renew my contract. Six months later I was diagnosed with breast cancer and that's up to the present day. Not much going on in my life, darling [laughs].'

What do we make of what appears as a disjuncture or apparent contradiction? Those who have developed mobility biographies as an approach expend considerable effort establishing the reliability of people's accounts. Yet here we want to focus on the structure of Laura's talk as a means to understand her travel and transport practices in the context of her everyday life. Thus we suggest her insistence on not being able to remember can be best understood, not as an evaluation of her own memory's reliability or validity, but rather as a statement about what is important in her life, and as an account of her own sense of temporality, which does not correspond with the linear, causal accounts of the existing mobility biographies literature. Laura also tells us, '… the only thing I can really remember is this last –, that's been very hard for me is at this last six, seven years, properly, because I've been through cancer.' In this condensed narrative, Laura refuses any kind of straightforward linear temporality; instead she suggests her own temporality – a continuous present of six or seven years of living with cancer – a present that has rendered the past, the time before cancer, almost invisible to her. In paying attention to her rhetorical strategies of repetition we suggest that concerns about reliability are misplaced. Accounts are not only about the past, but also about the ways in which the past is experienced in the present, which reveals a person's sense of time. For interpretive qualitative researchers using the interview method, these qualities can be seen as a resource rather than a stumbling block.

In contrast to the attention given to events and transitions in much of the extant mobilities biographies literature, this focus does not necessarily help with understanding Laura's account. A framework stressing events and transitions can make narrating a lived life challenging for those who don't have these or other conventional narrative hooks (Byrne 2003). Laura is not married, has no children or fixed employment, and does not have an easy employment or educational history to tell. Rather than being a one-off event, her cancer is an ongoing, unfolding story; one which calls into question the assumptions of linear time. Laura's own account of time is more truncated and condensed though, taking the form of a continuous present tense of six years and reminding us that lives are not always experienced or remembered as linear time.

This quality also means that it is difficult to trace any patterns of causality in Laura's account. Trying to understand what 'causes' Laura to be a walker or to not own a car is challenging. Although she is insistent about being a walker, her account is told through an apparently moral story about those who spend money needlessly on buses or are too lazy to walk and can therefore end up with diabetes or become obese. At the same time, it becomes apparent that she readily uses other people's cars, not only in the form of lifts from her father to her hospital appointments for chemotherapy, but also in the form of her boyfriend's car. She further recounts that if she had enough money she would actually buy a car herself, although in keeping with her narrative of appropriate spending it would be a Mini rather than a Jaguar. Such discussion helps to contextualize Laura's account of herself as a walker, both in terms of what appears to be financially feasible for her and the concomitant necessity of relying on others for car journeys.

While money, specifically the apparent lack of money, is dominant in Laura's account, she also makes reference to extensive foreign travel. Despite her insistence that she cannot afford a car, she does appear to be able to afford travel abroad (although again not too much, in contrast to her sister who goes abroad six times a year, which she views as 'a bit extreme'). It is not clear from her account whether her travel abroad is funded by some investments her father has made, and is paid for by him, rather than her own money. It is also not clear if her father is prepared to pay for travel abroad, but not to support Laura financially in other ways (for example, by buying a car). Whatever the case, there are neither straightforward economics nor recognizable causal relationships at play here. While Laura may be a walker on a day-to-day basis, a focus on this would miss the importance of foreign travel in both her mobility and personal imaginaries. Here we would fail to grasp much about how Laura understands and organizes her travel practices. Therefore, we suggest that the traditional life course-based mobility biographies would struggle to work with the 'data' of Laura's life. Indeed, it is difficult to see how conventional transport modelling can engage with Laura's life at all.

## Conclusion

In this chapter we have described the emergence of a mobility biographies approach in transport studies – a new, empirically-grounded method for understanding changing travel behaviour throughout people's lives. Yet, while welcoming this innovation, we have identified ways in which, as currently practiced, it risks sharing some of the limitations of existing transport studies research, despite its intention to address the dynamic nature of people's lives. Limitations of existing transport studies research include realism, linear time, the segmentation of people's lives into events and transitions, and the search for causality, among others. In seeking to extend the possibilities of mobility biographies we outlined an explicitly interpretative approach to people's narratives: one which demonstrates the limitations of reading off life events and transitions from people's accounts of their lives. We have also suggested that a key factor in understanding why conventional models fail is their inability to account for the complexity of people's lives. We have reviewed the contribution that narrative data from the first qualitative panel study focusing on transport and travel practices can make to an emerging literature on mobility biographies in the field of transport studies. This, we have argued, particularly when viewed through the lens of the interpretive social sciences, opens up the potential for a much richer perspective on biographical accounts of mobility as multiple, sometimes inconsistent, contradictory or non-linear 'trajectories'. Rooting the study of travel and transport behaviour in a dynamic empirical examination of the ways in which life stories and mobility histories are recounted, the meanings that are attached to everyday practices, and personal perceptions of time offer the prospect of a more nuanced understanding of how and why people live their lives. While the task of drawing on this revamped approach to mobility biographies in order to inform transport models and transport planning remains considerable, existing models will necessarily continue to fail unless we can design new approaches sensitive to and which can articulate more complex accounts of people's lived experience of travel. From within the policy process, there is an increasing challenge for transport planners and modellers to provide appropriate evidence to support the relevance of their models. We suggest that achieving any step change in transport behaviour will surely require an equivalent step change in the methods and models we use to understand people's lives.

## References

Aldred, R. and Jungnickel, K. 2012. Constructing mobile places between 'leisure' and 'transport': a case study of two group cycle rides. *Sociology*, 46 (3), 523–39.

Andrews, M., Sclater, S.C. and Squire, C. 2000. *Lines of Narrative: Psychosocial Perspectives*. London: Routledge.

Axhausen, K.W. 2008. Social networks, mobility biographies and travel: the survey challenges. *Environment and Planning B*, 35 (6), 981–96.

Büscher, M. and Urry, J. 2009. Mobile methods and the empirical. *European Journal of Social Theory*, 12 (1), 99–116.

Byrne, B. 2003. Reciting the self: narrative representations of the self in qualitative interviews. *Feminist Theory*, 4 (1), 29–49.

Chamberlayne, P., Bornat, J. and Weingraf, T. 2000. *The Turn to Biographical Methods in Social Science*. London: Routledge.

Clifton, K.J. and Handy, S.L. 2001. *Qualitative methods in travel behaviour research*. Paper to the International Conference on Transport Survey Quality and Innovation: Kruger National Park, South Africa.

Daniels, S. and Nash, C. 2004. Lifepaths: geography and biography. *Journal of Historical Geography*, 30 (3), 449–58.

Delbosc, A. 2012. The role of well-being in transport policy. *Transport Policy*, 23, 25–33.

Elliott, J., Miles, A., Parsons, S. and Savage, M. 2010. *The design and content of the 'Social participation' study: A qualitative sub-study conducted as part of the age 50 (2008) sweep of the National Child Development Study*. CLS Working Paper 2010/3. London: Centre for Longitudinal Studies.

Flyvbjerg, B., Holm, M. and Buhl, S. 2006. Inaccuracy in traffic forecasts. *Transp. Rev.*, 26 (1), 1–24.

Frändberg, L. 2008. Paths in transnational time-space: representing mobility biographies of young Swedes. *Geogr. Ann. B*, 90 (1), 17–28.

Freeman, M. 2010. *Hindsight: The Promise and Peril of Looking Backward.* Oxford: Oxford University Press.

Jain, J. and Lyons, G. 2008. The gift of travel time. *Journal of Transport Geography*, 16 (2), 81–9.

Lanzendorf, M. 2003. *Mobility Biographies. A New Perspective for Understanding Travel Behaviour*. Paper to the 10th International Conference on Travel Behaviour Research: Moving through Nets: The Physical and Social Dimensions of Travel, Lucerne, 10–15 August 2003.

Lanzendorf, M. 2010. Key events and their effect on mobility biographies: the case of childbirth. *International Journal of Sustainable Transportation*, 4 (5), 272–92.

Larsen, J., Urry, J. and Axhausen, K.W. 2006. *Mobilities, Networks, Geographies*. Aldershot: Ashgate.

Lyons. G. and Urry. J. 2005. Travel time use in the information age. *Transportation Research Part A*, 39 (2–3), 257–76.

Markovich, J. and Lucas, K. 2001. *The Social and Distributional Impacts of Transport: A Literature Review*. Working Paper N° 1055, Transport Studies Unit, University of Oxford.

Mason, J. 2002. *Qualitative Researching*. London: Sage.

Mason, M. 2010. Sample Size and Saturation in PhD Studies Using Qualitative Interviews. *Forum: Qualitative Social Research* [Online], 11 (3), Art. 8. Available at: http://www.qualitative-research.net/index.php/fqs/article/view/1428/3028 [accessed: 07 December 2012].

Mokhtarian, P., Salomon, I. and Redman, L. 2001. Understanding the demand for travel: it's not purely derived. *Innovation*, 14 (4), 355–80.

Parkhurst, G. and Shergold, I. 2009. *Literature review: the treatment of social and distributional impacts in appraisal and evaluation*. Project Report, Department for Transport, London.

Pas, E. 1990. Is travel demand analysis in the doldrums?, in *Developments in Dynamic and Activity-Based Approaches to Travel Analysis*, edited by P. Jones. Aldershot: Avebury, 3–27.

Pink, S. 2012. *Situating Everyday Life: Practices and Places*. London: Sage.

Pooley, C.G., Turnbull, J. and Adams, M. 2005. '… everywhere she went I had to tag along beside her': family, life course, and everyday mobility in England since the 1940s. *History of the Family*, 10 (2), 119–36.

Riessman, C.K. 1993. *Narrative Analysis*. London: Sage.

Riessman, C.K. 2002. Analysis of Personal Narratives, in *Handbook of Interview Research: Context and Method*, edited by J.F. Gubrium and J.A. Holstein. London: Sage, 695–710.

Roberts, B. 2002. *Biographical Research*. Buckingham: Open University Press.

Savage, M., Bagnall, G. and Longhurst, B. 2005. *Globalisation and Belonging*. London: Sage.

Scheiner, J. 2007. Mobility biographies: elements of a biographical theory of travel demand. *Erdkunde*, 61 (2), 161–73.

Taylor, S. 2012. 'One participant said …': the implications of quotations from biographical talk. *Qualitative Research*, 12 (4), 388–401.

Thomson, R. 2007. The qualitative longitudinal case history: practical, methodological and ethical reflections. *Social Policy and Society*, 6 (4), 571–82.

Timms, P. 2008. Transport models, philosophy and language. *Transportation*, 35 (3), 395–410.

Urry, J. 2000. *Sociology Beyond Societies: Mobilities for the Twenty First Century*. London: Routledge.

Urry, J. 2007. *Mobilities*. Cambridge: Polity.

# PART III
# Innovative Strategies

# Chapter 10
# MobiMart: Pilot Test for a Sustainable Mobility Credit System in Bologna

Katrin Dziekan, Giuseppe Liguori and Dora Ramazzotti

## Introduction

To create a sustainable transport system, various efforts in many fields of society and on different levels are necessary. One of the main challenges in this process is encouraging people to adapt their travel behaviour towards more sustainable modes. Two possible ways to do this include changing the available travel options and changing individual travel behaviour without altering travel options. Ideally, a combination of both would be most effective and would help to create a multimodal mind-set (Dziekan 2011), allowing people to be informed about different modal options while also knowing how to use them. Achieving a multimodal mind-set would conceivably mark a milestone in moving towards a sustainable transport system. Measures aiming to change travel behaviour are often described as 'push and pull' measures, where 'pull' measures make a given mode more attractive and 'push' measures make it less attractive or even prohibit it. Policy measures focusing on changing or reducing the demand for private car use are implemented under the term 'travel demand management' or 'mobility management' (Gärling and Fujii 2009).

Both practice and literature show that well-designed, incentive-based systems are very effective in changing consumer behaviour. Incentives are a widely used and researched type of measure (Laffont and Martimort 2002). Incentives are effective for consumer behaviour, but questions remain regarding whether they can also help to change travel behaviour as well as how the incentive system must be designed. What do we know about how to influence travel behaviour using incentives though? From psychology, we know that behavioural changes are only temporary if given a monetary pay off. Feedback on behaviour, such as whether or not a set target has been reached with a given behaviour, is a very important part of building up intrinsic motivation (Graham-Rowe et al. 2011). Thus, particular consideration should be given to designing the right level of incentives. Psychological research also shows that small incentives are often more effective when developing intrinsic motivations within the individual, while large incentives only strengthen external motivations that disappear when the incentives are no longer given (Schlag 2009).

The City of Bologna, together with SRM (the public transport authority), developed and tested an innovative sustainable mobility credit system called MobiMart, which was conducted as part of the CIVITAS MIMOSA project (see http://www.civitas.eu/mimosa). This is a project within the CIVITAS Initiative (http://www.civitas.eu), which is co-financed by the European Commission and helps cities to test and implement innovative solutions to achieve sustainable and vital cities. At the same time, it generates new knowledge that can be applied by other cities.

The original strategy for achieving more sustainable transport behaviour in Bologna was to implement one flexible pollution trading mechanism inspired by the emission trading mechanism included in the Kyoto Protocol. The flexible mechanism is based on voluntary emission reductions that can be traded and exchanged on a dedicated platform and bought by entities wanting to compensate for their emission levels. This is considered an attractive alternative to the European Union´s Emission Trading Scheme (EUETS), which was implemented in 2005 and is based on the idea of setting a defined level of allowed pollution. Companies that do not meet this level must then 'buy' credits rather than volunteering to take part.

The specific objective of the MobiMart measure, seen as a pull measure, is to shift trips made by motorized private transport to more sustainable transport options, such as flexible transport services, carpooling or cycling. The idea of the mobility credit mechanism is to reward positive (sustainable) behaviours related to transport and to convert saved $CO_2$ into 'mobility credits', which can then be traded for the benefit of the actor involved. The challenge is first to develop a valid methodology to certify the mobility credits (for example, by finding out how many kilometres people travel using sustainable modes instead of unsustainable modes and then devising a way to convert this into greenhouse gas savings). Besides facing the challenge of how to measure the mobility credits, a second important task was to identify effective incentives that stimulate intrinsic motivation, persuading people to change their travel behaviour. This chapter presents the key results from three pilot tests, all within the MobiMart project: a marketing campaign for a flexible transport service, a carpooling trial and campaign called 'Citizens by Bike'. The following section describes the CARTESIO network, whose guidelines were used to develop a mechanism for mobility credit certification. Sections 'Flexible Transport Services: a pilot test on colBUS – La Navetta del Borgo', 'Carpooling among Emilia-Romagna region employees' and 'Citizens by Bicycle' present the pilot projects and the main results. Conclusions and a general discussion of the results close this chapter.

## The CARTESIO network as a certification body

Italy has ratified the Kyoto Protocol and has set a national objective to reduce greenhouse gas (GHG) emissions to 6.5 per cent below 1990 levels. To achieve this, emissions in Italy in the period from 2008 to 2012 should not exceed 485.7 million tonnes of $CO_2$. On an annual basis, this objective corresponds to a saving of 95 million tonnes of $CO_2$. The reduction of 30 million tonnes of GHG emissions will be achieved by the industrial sector through the European Emission Trading Scheme (EU 2003). For the remaining 65 million tonnes, cooperation among government, regional authorities and local bodies was thought to be the most efficient way to reduce emissions. In addition to the promotion of eco-efficiency in industrial and private consumption, transport and buildings are the main sectors and movements engaged in this reduction. Regional authorities and local bodies additionally possess important competences related to these sectors. Some local authorities have already taken action through the adoption of voluntary tools aimed at the accounting, planning, communication and valorization of reduction and compensation actions.

The CARTESIO network is committed to surveying and analysing the possible contribution of regional authorities and local bodies to achieving the Kyoto targets and to identifying commonly adopted methods of reporting and quantifying results. It is promoted by six Italian regional authorities: Emilia-Romagna, Lazio, Liguria, Lombardy, Tuscany and Sardinia, and is open to public and private actors. In 2012, the CARTESIO network involved more than 180 organizations from 17 Italian regions.

The network aims to reach and promote collective solutions for the sustainable management of both industrial and urban areas, with an approach aimed at improving existing synergies.

The scope of areas in CARTESIO is very broad and includes cluster approaches for EMAS (Eco-Management and Audit Scheme), eco-industrial parks, product supply chain policies, governance and climate change. Until 2010, there were no specific guidelines and transfer procedures established for behaviour change from unsustainable modes (for example petrol-driven cars) to sustainable modes (for example bikes). Therefore, the CARTESIO network welcomed the idea of the MobiMart project. Based on the general CARTESIO guidelines (Rete CARTESIO 2010), new certification procedures were developed, together with the MobiMart partners, and tested in the three pilot activities. Supervision and authentication activities were conducted during the testing phase by an external academic auditor to confirm the MobiMart project's compliance with the CARTESIO guidelines and to both identify and certify the experimental tests as 'GHG reduction projects' as defined by the guidelines. The outcome of this process is the possible validation of corresponding 'reduction quotas', which can be entered into a national register.

Supervision and authentication activities were primarily focused on emission reduction objectives, general eligibility requirements (do the pilots produce greenhouse gas emission savings?) and baseline referencing and monitoring. MobiMart consisted of three pilot tests (see Figure 10.1). It was managed by SRM and also involved the public transport company of Bologna, the regional administration, Bologna citizens and other market players. The pilot tests and results are described in the next sections.

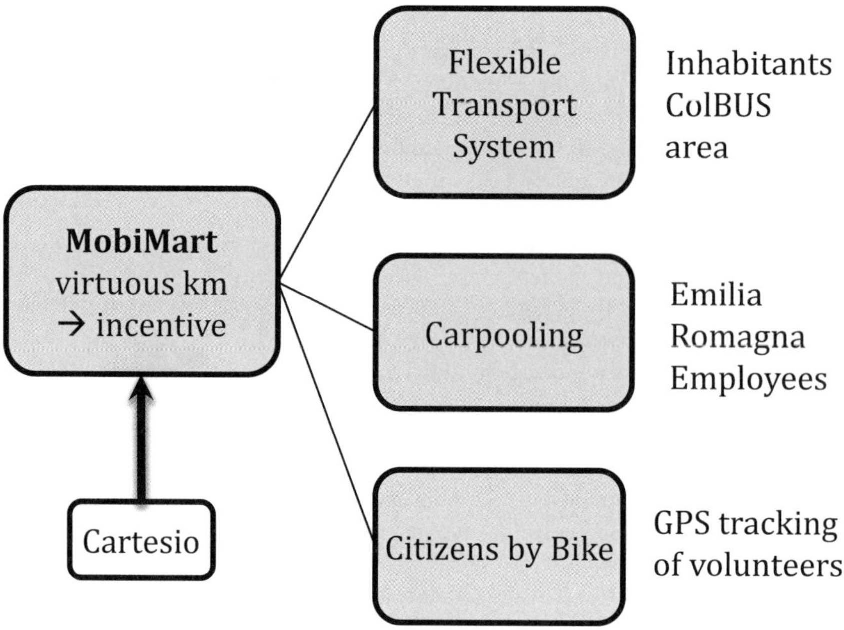


**Figure 10.1 Overview MobiMart pilot actions**

## Flexible Transport Services: a pilot test on colBUS – La Navetta del Borgo

The first pilot test in the MobiMart project was colBUS (La Navetta del Borgo), an experimental urban public transport line service that was in operation in a neighbourhood district in Bologna from June 2009 to August 2011. According to the Bologna Urban Traffic Master Plan (Municipality of Bologna 2006) approved by the Municipality of Bologna in 2007, an area was chosen that had not yet had an urban bus service and where a demand was recognized. In this area the distribution of the 12,000 inhabitants was not homogeneous; the part developed along the Emilia road is greatly populated while a second zone is sparsely populated. One bus was operating with yearly costs of 35,000 Euro for SRM. ColBUS could be booked by phone in advance and allowed customers to move within the district or to be linked to the main lines of public transport. This system is considered a Flexible Transport Service (FTS) and was operated according to itineraries and schedules agreed to at the time of booking between the user and the operator on the phone.

The FTS-bus had already been operating for almost one and a half years when the pilot test campaign within MobiMart, entitled 'Fall in love with colBUS – La navetta del Borgo', was launched in January 2011 by SRM. On the week of Valentine's Day (from the 14th to the 19th of February in 2011), colBUS was available for free to all citizens. The aim of the campaign was to raise awareness of the colBUS service by inviting citizens to take a free trial; the expected and hopeful outcome was to increase their use of public transport. The pilot week was announced through a two-week bulk advertising campaign that included the direct mailing of information to the 199 respondents (those who provided personal data during a survey of citizens and potential bus users conducted in 2010 in the district). In addition, flyers were posted on public bulletin boards around the Borgo Panigale District and at the most popular colBUS stops as well as virtually on the district website.

As a further incentive, those who used the bus during the campaign week immediately qualified for a lottery to win one free colBUS monthly ticket. The passenger numbers were counted and reported from the start of the service in June 2009 onwards. While the average trips per colBUS before the campaign stood at about 42, the passenger numbers per week in Figure 10.2 show that during the campaign, the number increased to 80 passengers per week.

An initial increase in passenger numbers was already noticed when the dissemination campaign started at the end of January. The average number of passengers using the colBUS service per month in 2010 was 204; results were similar in January 2011, when 212 passengers were recorded. With the help of the pilot campaign and related promotional campaigns, 326 passengers used colBUS in February 2011, increasing the average number of passengers per day from 9 to 14.

The number of colBUS passengers during the campaign week was higher than the previous average (see Figure 10.2). Additionally, during the campaign, the average number of passengers was higher in comparison to the Business as Usual (BaU) scenario calculated from the 'before data'. The BaU data was based on passenger numbers between June 2009 and January 2011 and was adjusted to seasonal data (black scattered data and trend line in Figure 10.2) as colBUS is affected by seasonal fluctuations in travel behaviour.

The key to promoting a shift from individual mobility to collective mobility is communication. In fact, we found that the number of passengers per day had already begun to increase during the advertising campaign A total increase of 411 single trips was registered during the 29 weeks observed.

In order to quantify $CO_2$ reductions by the campaign week, travel behaviour before and after the campaign week were compared. The data was obtained from telephone interviews conducted with people who used the colBus service (n=20). Interview questions were formulated to collect information on travel behaviour before, during and after the campaign week. Ownership and access to cars and motorbikes, driver's license possession, reason(s) for using the colBUS as well as satisfaction levels with the service were collected. The number of passengers and number of trips per passenger were revealed from weekly booking data available from the operator.

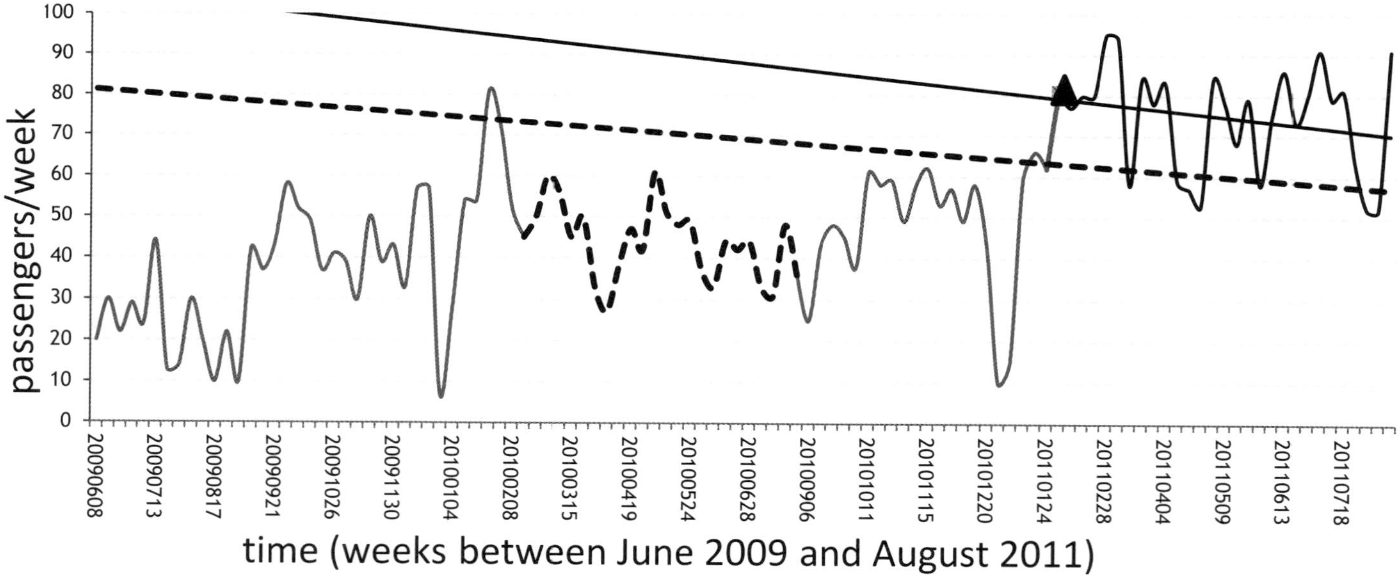


**Figure 10.2** Number of passengers per week on colBUS between June 2009 and August 2011 – campaign week (February 14th–19th as indicated with a black triangle) – corridor: indicated business-as-usual scenario derived from data from previous year

Emission factors are presented in Table 10.1 and take into account all participants (n=20) and all trips made during the pilot week as well as their trips made before (reported in 'before data' collection).

**Table 10.1 Average emissions factor of colBUS users in pilot test**

| | $CO_2$ (in g) | $NO_x$ (in g) | $PM_{10}$ (in g) |
|---|---|---|---|
| Average colBUS single trip emission (average mileage) | 733 | 1.75 | 0.39 |
| Average private vehicles single trip emission (average mileage) | 1 168 | 1.02 | 0.20 |
| **Difference** | **–435** | **0.73** | **0.19** |
| **ColBUS pilot impact (on 411 extra single trips replacing private car usage)** | **–178 923** | **298.76** | **77.03** |

In order to assess the impact, the emission reduction or increase per trip must be multiplied by the increased number of passengers (single trip) resulting from the pilot week (411 trips). Results are positive in terms of $CO_2$ reductions (179 kg saved), but are negative in terms of $NO_x$ (298 g more) and $PM_{10}$ (77 g more) emissions (see Table 10.1). These results could be improved by using a bus with a higher EURO-norm classification.

The colBus pilot was successful in increasing colBUS use and reducing $CO_2$ emissions. The related certification was requested from external academic auditors. Regarding the incentives, it was shown that a free trial period for passengers could attract users to the FTS service.

## Carpooling among Emilia-Romagna region employees

Carpooling was chosen as the second pilot. The term 'carpooling' refers to several people using one private car to make a trip together.

For the carpooling pilot, the participation of a large public body – the Emilia-Romagna Region Administration (approximately 800 employees) – was arranged. In order to identify available volunteers, SRM and the Emilia-Romagna Region Administration Mobility Manager conducted an information campaign among the employees using the intranet, a road show and an already activated website that promotes the formation of carpooling groups.

A specific on-board device (AZregolo) and a microchip card were provided to collect data on shared trips. These served to give an exact calculation of $CO_2$ savings by the algorithm described later. For each group, the on-board device collected all the needed data for MobiMart carpooling purposes using the smart card function. The first passenger who checked in was considered the driver; thus, the car being used was that associated with the driver and their profile. The others who checked in were automatically passengers. At the end of the trip, each passenger checked out, so as to allow a precise registration of kilometres travelled on board by each passenger.

At the end of each month, the server calculated the amount of 'virtuous' (saved kilometres per carpooling team) kilometres for each group and for each subscriber. The system was also able to calculate the amount of money that each driver should charge, and could even process the payment.

In total, 39 volunteers were recruited, of which eight were willing to participate as passengers and five as drivers, with the final 26 having applied for both roles. After a detailed analysis of potential paths/groups by SRM, 24 potential users in nine groups were selected. Three groups participated in the six-month pilot test that started on 24 April 2011.

The first direct incentive for participation was the savings incurred by each individual participating in a group that implemented the carpool system. A second incentive was the free reserved parking for each group of carpoolers close to their workplace. As far as prizes were concerned, due to the fact that many of the potential carpoolers were commuters travelling from other municipalities or counties, mobility-related rewards (such as urban bus tickets) were considered ineffective and difficult to exploit. Furthermore, the Emilia-Romagna Region already grants its employees mobility-related benefits. Therefore, to be attractive, another kind of prize was explored, and consequently 'leisure packages' were secured as prizes for the most virtuous car-poolers (the team that saved most kilometres). In addition, a lottery prize was available as a further package.

In order to assign the prizes, the saved kilometres per carpooling team were calculated. From this, $CO_2$ savings were calculated, along with the quantity of emissions emitted (and avoided) per car. Data on car characteristics were available, since every user was logged into the database with its car type. Table 10.2 shows the three teams in the pilot: Team 1 was composed of two members who owned respectively one car and one motorbike; Team 2 was made up of two persons who both had a car; and Team 3 was made up of two persons, only one of whom had a car. Team 3 was admitted to the test even though they did not have an alternative vehicle. In the phase of calculation of $CO_2$ emission reduction, the benefit obtained by the team was considered as zero.

The calculation of the baseline and the calculation of the carpooling tests were done on the basis of data released by the participants and by AZregolo using COPERT IV (www.emisia.com/copert/) methodology. It was assumed that if the AZregolo data did not record a trip, the person was using an alternative vehicle outside the carpooling scheme.

The $CO_2$ emission baseline was calculated by multiplying specific emission factors in g/km and the distances driven. Emissions were summed up for all participants in the baseline and in the project phase. Emission reductions were determined by comparing total emissions in the baseline with total emissions in the project phase.

**Table 10.2** Calculation example for three carpooling teams of $CO_2$ savings for the carpooling test period

| | Team 1 | | | | Team 2 | | | | Team 3 | |
|---|---|---|---|---|---|---|---|---|---|---|
| | Baseline | | Test | | Baseline | | Test | | Baseline | Test |
| **Vehicle** | A1 | B1 | A1 | B1 | A2 | B2 | A2 | B2 | A3 | A3 |
| **Number of days in test** | 132 | 132 | 132 | 132 | 132 | 132 | 132 | 132 | 132 | 132 |
| **Number of days in carpooling** | 0 | 0 | 49 | 49 | 0 | 0 | 18 | 38 | 0 | 6 |
| **Number of pooling trips (one way)** | 0 | 0 | 70 | 70 | 0 | 0 | 75 | 75 | 0 | 8 |
| **Total distance per vehicle (km)** | 2 350 | 2 350 | 2 350 | 2 350 | 14 124 | 13 675 | 14 124 | 13 675 | 1 822 | 1 822 |
| **Distance without carpooling (km)** | 0 | 0 | 0 | 603 | 0 | 0 | 1 222 | 2 121 | 0 | 0 |
| **Distance using carpooling (km)** | 2 350 | 2 350 | 2 350 | 1 747 | 14 124 | 13 675 | 12 902 | 11 554 | 1 822 | 1 822 |
| **Vehicle use coefficiency** | 1 | 1 | 1 | 0.74 | 1 | 1 | 0.91 | 0.84 | 1 | 1 |
| **$CO_2$ total emission (t)** | 0.51 | | 0.47 | | 3.99 | | 3.51 | | 0.27 | 0.27 |
| **Saving on $CO_2$ emissions (t)** | 0.0 | | **0.04** | | 0.0 | | **0.48** | | 0.0 | **0.00** |
| **Reduction percentage compared to baseline** | | | **8.5%** | | | | **12.0%** | | | **0.0%** |

Table 10.2 shows the calculation results of the three teams. It includes both the baseline, which was theoretically derived based on the information on available cars and kilometres from home to work, and the actual behaviour in the test period.

It was necessary to adjust and elaborate the raw data before final analysis, especially in order to exclude obvious errors such as invalid log-ins or missed log-outs by smart card users. The results of the six-month test period for three groups (two persons in each group) were promising, since they had saved roughly 0.52 tons of $CO_2$. This is approximately a reduction of 11 per cent compared to the emissions before the test (see Table 10.2).

Apart from the direct impact, the evaluation process revealed one important lesson regarding how to best to launch a successful carpooling project: it was important to involve a large company (ideally a local governmental body) that has a large number of employees not working in shifts. Therefore, work shifts should be considered as a barrier to matching transport needs. The first MobiMart carpooling attempt involving a local health services company failed because work shifts hindered the start of the pilot. From stakeholder interviews in the evaluation process of the MobiMart measure it became clear that another key factor was the availability of parking spaces in close proximity to the workspace. The lack of parking and high fees in parking garages make reserved spots for carpoolers in close proximity to the building a strong incentive.

The distance of the employee from the place of employment should be meaningful enough to produce an economic incentive, which is a real force to involve the carpoolers, as they will accept restrictions in their freedom of movement in order to coordinate trips with other people. Another important success factor is the availability of an alternative plan in case the driver cannot provide the return trip. This can be, for instance, an option to use the company fleet or to have a rental car at one's disposal.

To summarize, the carpooling pilot test was implemented successfully, and the emission saving calculation was certified by CARTESIO. However, the attractiveness for employees was not high enough. The decision was made not to repeat the pilot after the first testing period or to test it with other partners.

## Citizens by Bicycle

Citizens by Bicycle, the third MobiMart pilot test, was an initiative organized by SRM with the aim of encouraging the use of bicycles and of raising awareness about the topic of sustainable mobility among citizens. To attract people to participate, a challenge was launched. Who could substitute the most kilometres driven by car with kilometres by bike within a given time frame? A baseline – which was unbiased, because the challenge idea had not been launched yet – was established with a questionnaire during European Mobility Week 2010 in Bologna and other related events. Three hundred and sixty-two of the 746 people interviewed showed interest in participating in the pilot. In this initial questionnaire, the respondents were asked to provide an average weekly overview of their trips, type of trips and transport mode.

On 23 May 2011 the campaign itself kicked off by announcing the challenge. It was preceded by a series of awareness-raising campaigns through press releases and social networks. Participants were asked to record their movement for two weeks by GPS-tracking via a website as well as in a special mobility diary.

For the GPS tracking, cooperation was arranged with Endomondo.com, a site specializing in GPS tracking for sports activities. Endomondo provided SRM with a site specifically created for the Bologna 'Citizens by Bicycle' campaign (http://www.endomondo.com/campaign/mimosa). The website is available to download for free as a mobile application for tracking cycling trips with GPS technology.

Incentives such as cycling-related prizes for the winners of the challenge and a lottery with prizes for participants were also planned.

**Figure 10.3 Campaign logo of the bicycle pilot**

*Source:* SRM

The pilot results in terms of emission reductions are based on the group of pilot participants that completed both the before paper questionnaire and the travel diary (n=138).

The aim of the pilot was to substitute private motorized means of transport with bicycle trips. The analysis of the survey data before and during the pilot test showed that the use of bicycles decreased. The main reason for this unintended result was a bias in the data caused by different data collection methods. Whereas participants reported their average travelling habits (retrospective self-reported behaviour) in the 'before' survey, the pilot test data consisted of actual logged bicycle trips (actual behaviour). Further, a weather bias should be taken into account, considering the amount of precipitation in June 2011 (the pilot started on 23 May and ended on 19 June). During this period it was well above average, with 146 mm of rain compared to an average of 57 mm per month. This huge amount of rain probably affected modal choice in favour of car use.

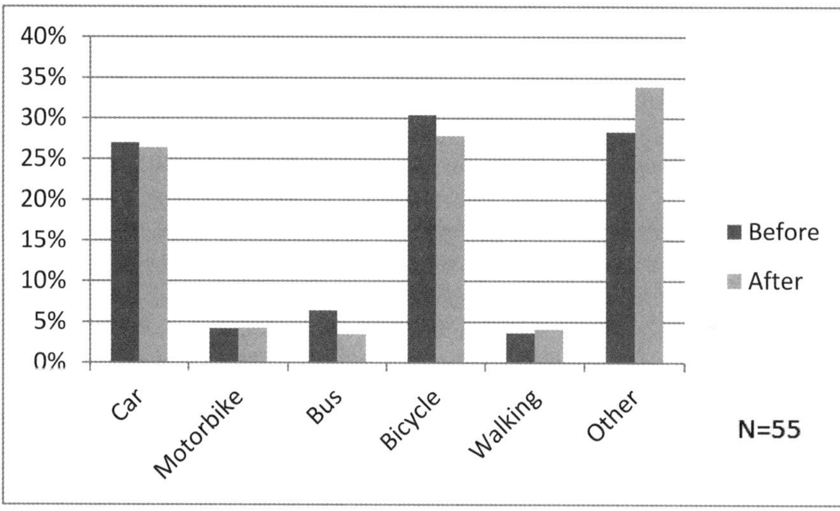


**Figure 10.4 Self-reported modal split (without leisure trips) of participants (N=55) in the 'citizen by bike' pilot**

Data were collected through a before and after survey; in total, 55 people responded to both surveys. Figure 10.4 shows aggregated data from the travel diaries (excluding leisure trips). When assessing the individual behaviour of participants, one group (n=23) increased cycling, while a smaller subgroup (n=10) at the same time reduced motorized kilometres driven (a modal shift generating impacts on $CO_2$ emissions). Calculette, a standard instrument for calculating $CO_2$ footprints (http://www.ademe.fr), was adapted to meet local needs in order to calculate $CO_2$ equivalent emissions saved. Combined before/after transport footprints were calculated for each participant, based on 0.302 kg of $CO_2$ issued per kilometre by car and 0.088 kg of $CO_2$ per kilometre by motorbike.

Based on these footprints and individual modal shifts, the calculation of $CO_2$ equivalent savings was positive: 41 kg were saved during the pilot week by the whole group of 55 participants, with a potential saving of 2.13 tonnes per year if the behaviour remained constant (52 weeks).

The evaluation process (which included interviews with stakeholders, among other methods) showed that the main reason participants took part in the pilot action was not to simply win a prize in the lottery. Personal motivation and a shared objective of reducing $CO_2$ emissions by all participants were more important. Also the 'challenge mechanism', which was supported by social networks on the Internet, worked very well; 11 teams were created, representing different areas within the city, and their kilometres driven were updated so that progress in the competition could be monitored by participants in real-time. A final event in Bologna's main square allowed all participants to meet. One bicycle was awarded to the person who substituted the most motorized trips by bicycle and another one was handed out to a randomly selected participant. In addition, an assortment of small prizes was raffled off among all participants.

To summarize, this pilot was a success, and in particular, the 'challenge' component emerged as a very effective incentive. Elements of this pilot were used for the European-wide 'Bike Challenge' that was successfully conducted in May 2012 and will be repeated in 2013 (http://www.srmbologna.it/ecc2013).

## Conclusions

The MobiMart idea was tested via three pilot tests: a marketing campaign for a flexible transport service, a carpooling trial and the 'Citizens by Bike' campaign. All pilots succeeded in certifying saved GHG emissions with the use of CARTESIO. Thus, the first step towards a mobility credit system was achieved. It is very complex to calculate GHG emission savings from changes in mobility behaviour. Nonetheless, the three pilots served as a first step towards the certification of GHG savings through sustainable mobility measures, which is a promising development. Nevertheless, the MobiMart system – a trading mechanism with mobility credits – still requires development and refinement.

One specific issue of interest relates to the measurement methodology. Based on the pilot, it is now clear that measurements intended for large industrial plants cannot be applied directly to the mobility sector since real-time measurement requires a relevant amount of resources in terms of time and money. A new methodology must be applied, based upon local (detailed) statistics and models on car use as well as average pollution per kilometre per inhabitants. An urban cyclist, for example, saves $CO_2$ by just cycling instead of driving; this is without the need to demonstrate saved GHG emissions through a before/after analysis. In this way, even small groups or individuals can measure GHG emission reductions, and thus apply the MobiMart conversion methodology.

Once the FTS service contract expired it was not renewed due to financial problems. A second round of the carpooling pilot was not launched due to the lack of interest from the partner and the potential carpoolers. The biking pilot though has been very successful. Based on this success, a new innovative idea has emerged from the test. The Internet-based GPS tool for the reporting of bicycle trips tested and adapted in the MobiMart project has been utilized since May 2012 in the *Biking Challenge* with participating cities across Europe.

Did the incentives influence the travel behaviour of citizens in Bologna, moving them towards adopting more sustainable transport modes? The analysis of the different steps taken in the pilots has given us the impression that incentives are good, but that they alone will not change behaviour. Many citizens stated their interest and availability regarding involvement in the first phase, but in the end, only a few accepted the challenge to try to change their behaviour. We suspect this is due to the fact that the mobility options people had at their disposal remained unchanged. If some traffic restricting measures (push measures) had been implemented by local decision-makers or mobility managers (such as parking place reductions/parking cost increases, road pricing measures, increase of pedestrian areas, and so on), the results could have made a much greater impact.

So in the end, our answer is yes, incentives work for supporting behaviour change towards more sustainable transport modes, but they must be accompanied by strong promotional and advertising campaigns in order to share a 'vision' with pilot volunteers. Also, the 'challenge' mechanism launched was very stimulating and allowed the participants to feel as if they were part of a community in action that was doing something good for sustainable mobility. In this case the incentive mechanism was not a prize-based one, but a motivation one. This is indeed better because its impact endures beyond the end of the pilot project given that it relates strictly to the behaviour of the participant and not to the prize itself. Another key point is that a challenge mechanism can be cheaper than a prize-giving methodology.

The mobility credit idea was very innovative, both in Bologna and at the European level. But to what degree are private and public mobility managers ready to bet on it? Once the mobility credit methodology has been certified by official organizations, the success of this methodology will depend on the active participation of people and stakeholders in the mobility credit program.

**Acknowledgement**

Work for this chapter was co-financed by CIVITAS MIMOSA and by the European Commission – grant TREN/FP7TR/218953.

## References

Dziekan, K. 2011. Public Transport (Öffentlicher Verkehr), in *Verkehrspolitik. Eine interdisziplinäre Einführung*, edited by O. Schwedes. Wiesbaden: VS Verlag, 317–40.

EU. 2003. Directive ETS CE/87/2003.

Gärling, T. and Fujii, S. 2009. Travel behavior modification: theory, methods, and programs, in *The Expanding Sphere of Travel Behaviour Research. Selected Papers from the 11th International Conference on Travel Behaviour Research*, edited by R. Kitamura, et al. Bingley, UK: Emerald, 98–128.

Graham-Rowe, E., Skippon, S., Gardner, B. and Abraham, C. 2011. Can we reduce car use, if so, how? A review of available evidence. *Transportation Research Part A: Policy and Practice*, 45 (5), 401–18.

Laffont, J. and Martimort, D. 2002. *The Theory of Incentives: The Principal-Agent Model*. Princeton: Princeton University Press.

Municipality of Bologna 2006. Urban Traffic Master Plan (*Piano Generale del Traffico Urbano* – PGTU) [Online: Municipality of Bologna]. Available at: http://www.comune.bologna.it/trasporti/servizi/2:4036/4266/ [accessed: 05 December 2012].

Rete CARTESIO 2010. *Linee Guida per la definizione e attuazione di una strategia di riduzione delle emissioni di gas serra da parte delle pubbliche amministrazioni.* Available at: http://www.retecartesio.it/Documenti/LLGG_GHG_febbraio2010.pdf [accessed: 05 December 2012].

Schlag, B. 2009. Achievement and learning motivation (*Lern- und Leistungsmotivation*). 3rd edition. Wiesbaden: VS Verlag.

## Internet sources

http://www.civitas.eu [accessed: 27 September 2012].

http://www.civitas.eu/mimosa [accessed: 27 September 2012].

http://www.emisia.com/copert/ [accessed: 27 August 2012].

http://www.endomondo.com/campaign/mimosa [accessed: 20 September 2011].

http://www.ademe.fr [accessed: 18 May 2012].

http://www.srmbologna.it/ecc2013 [accessed: 28 September 2012].

# Chapter 11
# Status Quo and Perspectives for CarSharing Systems: The Example of *DriveNow*

Johanna Kopp, Regine Gerike and Kay W. Axhausen

## Introduction: evolution of CarSharing schemes

CarSharing is known as the professionally organized sharing of one or more vehicles by several users. The vehicles are not privately owned, but instead belong to the CarSharing organization (Millard-Ball et al. 2005). The idea of shared car-use originally emerged from self-organized, informal initiatives started by neighborhood communities. Sharing cars between neighbors or acquaintances shows a renaissance as a sort of private CarSharing but no longer falls strictly under the term CarSharing; the essential characteristic of CarSharing is the commercial nature of the offers presented by CarSharing organizations (Sakhdari 2006). The first CarSharing organization *ShareCom*, following the current understanding of the term, was started in Zurich in 1987. This was followed in 1988 by the first German CarSharing organization *STATTAUTO* in Berlin (Petersen 1995). The roots of this can be traced to what, at that time, was a new social orientation: the motivation to reduce costs and, more importantly, to reduce the fleet of cars caused by the congestion of the transport system, as well as the concept of sustainable and ecological forms of transport. It is therefore the purpose of CarSharing to enable a more efficient usage of available vehicles and to provide people with access to individual mobility in order to cover certain needs, despite to their decision to live without owning a car (Shaheen and Cohen 2007, Koss 2002, Wilke 2009).

New potential analyses and acceptance studies in the field of CarSharing and car elimination show that a user-friendly CarSharing offer can lead to complete elimination of car-ownership in private households as well as satisfying and compensating the needs of owning a second or third car (Shaheen 1999, Harms 2003, Steding et al. 2004). However, up to now, the available CarSharing models have not been convenient enough in comparison to owning a personal car, inasmuch as they are not able to induce the elimination of private cars in households (Fliegner 2002). Due to this high comfort deficit from a user perspective, CarSharing has been a niche product in Germany. In 2011 only 0.16 per cent of all driver's licence owners are member of a CarSharing organization (BCS 2011).

Therefore it is necessary for CarSharing organizations to offer an attractive way of satisfying mobility needs in order to be able to induce such an effect. As such, it has been possible to observe a further development in CarSharing models in the past years. Several institutions have recognized both this deficit and the high potential for mobilizing new CarSharing users with more flexible solutions. Major car companies such as VW (*Quicar*), Daimler (*Car2Go*) and BMW (*DriveNow*) have started CarSharing systems in several cities. As a result of a change in how people in cities – especially young people – relate to cars and due to the concurrent increase in the costs of keeping a vehicle, automobile manufacturers have begun to develop mobility solutions in addition to their core business of producing cars. Of these, *DriveNow* and *Car2Go* offer the greatest possible flexibility, which is commonly described in literature as a key component to transitioning from the current CarSharing niche and thus enabling CarSharing to expand and appeal to a wider set of users (Franke 2001).

The objective of this chapter is to provide an overview of how CarSharing schemes have developed over the last 30 years, from station-based models to free-floating systems. It also aims to describe how the member and usage structure has developed through changes in the CarSharing systems. In order to complete this overview, a literature research was conducted as well as a comparison of the results from studies on station-based CarSharing models with a usage structure analysis of the German free-floating Carsharing system *DriveNow*.

## Main characteristics of station-based and free-floating CarSharing systems

The following chapter describes the evolution of CarSharing schemes from traditional station-based systems to more flexible free-floating CarSharing models, with respect to various operating and the following usage aspects: legal form, spatial spread, registration and membership, vehicle access, vehicle selection, stations and parking areas, reservation and booking, driving relationships and return as well as costs and billing.

### *Legal form: professionalisation of CarSharing schemes*

In the 1990s, the majority of CarSharing organizations in Germany had the legal structure of a registered association (e.V.) (Rösch 2001). In addition to the convenience that this offered for the operators, it showed the social impetus behind its structure – members are allowed, and to an extent, forced to participate in decision-making. Furthermore, users and members are considered identical and have equal rights. As such, these organizations did not see other providers as competitors and developed into rather highly differentiated, small-scale operators.

Starting in the mid-1990s, cooperative CarSharing began to change this structure. Some of the then current providers began to reorganize themselves into corporations. Additionally, with the increasing size of the CarSharing organizations it was often necessary to transform their legal structure into a GmbH (limited company). This might be because after a certain point, the advantages of a GmbH outweighed those of a registered association. Moreover, after a CarSharing organization reached a certain number of users, it is no longer allowed to register as the association due to the anonymity that emerges (Rösch 2001, Pesch 1996).

Today, in contrast to Switzerland, where a national provider (Mobility) exists, the German market is characterized by a large diversity of institutions that have made different progress. There are still organizations that are dependent on voluntary work, whereas others see it as a strong and successful business model and therefore approach it professionally (Nobis 2006: 3). The goals of the company are often closely intertwined with the personal and political goals of the stakeholders of station-based CarSharing schemes. In recent years, we have seen an emerging harmonization and cooperation among CarSharing services. Nevertheless, a diversity of CarSharing organizations is still noticeable in Germany, where about 90 per cent of the users are divided among the three biggest cooperation systems (*stadtmobil*, *Cambio* and CarSharing from German Rail [Franchise]) (BCS 2012).

*Spatial spread of CarSharing schemes*

One can trace back the roots of CarSharing to Europe, but since the 1990s, CarSharing has spread to North America and Asia. On the global scale today, CarSharing exists in about 600 cities, 18 countries and on four continents with growth rates that have been recorded all over the world. In 2008, almost 348,000 members and 11.700 vehicles were registered (Shaheen and Cohen 2007). It is assumed that CarSharing vehicles are used by more than one person for most of the trips. In Germany, CarSharing tends to be concentrated in big cities with more than 200,000 citizens; further, it is mostly available in cities with populations of 100,000–200,000 inhabitants (Nobis 2006). In Germany, there are currently about 110 CarSharing providers serving the demand of 158,000 clients in 285 cities, with at total number of 4,600 vehicles at 2,200 CarSharing stations (BCS 2011).

*Registration and membership*

The prerequisite for the use of CarSharing is membership in the form of a contract with a CarSharing organization. In most cases, a registration fee is also required (Johnsen 2007: 9). With traditional and station-based CarSharing systems, a deposit has to be paid in addition to the fixed monthly subscription (Harms 2003). With the new free-floating CarSharing systems though, there are no deposits and fixed costs, only a one-time registration fee. In sum, members do not pay any fixed costs as the billing is strictly usage-dependent.

*Vehicle access*

In all CarSharing organizations, it is necessary to guarantee that members have continuous and independent access to the vehicles. Previously, access to a vehicle was facilitated by the transfer of keys in safes or vaults (Sakhdari 2006); members were rarely in possession of the keys or had to pick them up from offices (BASt

2004). However, this kind of key-transfer only worked for small organizations with a manageable numbers of users. Nowadays, access to a vehicle is facilitated though the use of chip cards in combination with electronic systems (via RFID) in all kinds of CarSharing schemes, both station-based and free-floating. Members receive their chip cards from the CarSharing organization after registration (TCRP 2005). In Germany, the owner of a vehicle (usually the CarSharing organization) has to guarantee that the user has a valid driver's licence at the moment of driving. As a result, some providers have started to stick obligatory RFID chips onto the driver's licence to guarantee that the user always has a valid licence when driving. Furthermore, the vehicles in the CarSharing fleet are equipped with appropriate reading devices. These are usually fastened to the windshield and even give information as to the status of the vehicle, including whether it is available, reserved or booked.

*Vehicle selection*

CarSharing schemes generally provide various models and categories of cars in order to comprehensively cover the diverse needs of users. The fleets include light-weight cars, middle-range cars and transporters. This trend of offering users a variety of vehicles has been widely maintained. Despite this though, some of the new CarSharing providers only offer one car model (such as *Car2Go*) or different vehicles of the same brand (such as *DriveNow*, *Quicar*). In regards to car maintenance, smaller organizations draw on the help of volunteers, while larger fleets have their own established car service or have external service providers for maintenance (BASt 2004). Some CarSharing organizations even offer their members incentives to get them to take on part in vehicle maintenance (for example, free minutes for refuelling).

*Stations and parking areas*

In the past, station-based CarSharing vehicles were not available in public street parking spaces and instead only available at fixed rented stations. These stations were located in residential areas as well as at the main entry points of public transport (such as train stations or stops) (Johnsen 2007, Shaheen and Cohen 2007). This approach meant that the user had to pick up the vehicle at a certain station and to generally return it to the same station. As a result of the positive effects of CarSharing, some of the German federal states reserved a number of street parking spaces for CarSharing. However, the lack of any national legal baseline today has led to isolated local solutions being implemented and based partly on surface reclassifications. The starting point for this was the consideration of locating CarSharing vehicles close to residential areas in order to increase user acceptance.

As a result, a neighbourhood-based system also developed, in which vehicles have residential parking permits for a particular licence area and are allowed to park anywhere within this area (such as *Flinkster* from German Rail or *zebra mobil*).

In contrast to CarSharing schemes with fixed stations, free-floating schemes think of each vehicle as an independent station, where vehicles can be parked in any free on-street parking space within a defined zone (for example, downtown within a big city). As a result, vehicles can be located closer to their trip origin or destination. Disadvantages can arise from the difficulties of finding a parking space due to the lack of specially designated CarSharing parking spaces. In general though, the development of a tight parking space supply represents a limiting factor for the spread of all CarSharing schemes (ITS 2008).

*Reservations and booking*

In the past, station-based CarSharing vehicles usually had to be booked a long time in advance and return-time also had to be determined from the start. In addition, a booking could incur heavy fines if the vehicle was returned late, due to the fact that the following users would not be able to find their booked vehicle in a specified location. Over time and through the use of computerized vehicle management though, station-based CarSharing organizations have been able to ensure that the only reservation necessary is the type of vehicle. The actual distribution of vehicles was then carried out spontaneously by the operator. In addition, return time could be made more flexible as the reservation period could be extended spontaneously depending on demand and availability.

In the course of further flexibilization through free-floating CarSharing, it became standard for vehicles to be booked and used on demand. Today, reservation is now possible 24 hours per day and can be done either by telephone or (most often) via the internet. Furthermore, it's possible to reserve a vehicle via smartphone applications (BMW Group 2011, TCRP 2005). A key improvement in recent time has been the simplification of the booking process. 'Instant Access' and 'Instant Booking' allow the user to access the vehicle without having to mandatorily reserve it. As a result, the utilization and attraction of CarSharing vehicles have substantially increased (Canzler 2010). An important element is 'Open End', the freedom to leave the return time open and to not have to specify the beginning and end of use beforehand.

Despite the fact that it is possible to spontaneously book vehicles these days, many station-based CarSharing schemes still require the return time to be set in advance. This allows the provider to rent out the vehicles in advance for longer periods of time (such as for months).

*Driving relationships and return*

In the traditional, station-based CarSharing systems, a vehicle can generally only be picked up from one point and then returned to that same point (A to A). Over time, there has been some flexibilization: it is partly possible to return a vehicle to a second station or to an entire quarter serving as a station (A to B, One Way). In this context, the concept of 'One Way' allows for the offset of the spatial limitations: users can decide for themselves where (and at what station) they park the vehicle. In simpler terms, the vehicle does not have to be returned to the starting point. At the beginning, this was hardly possible due to the fact that the respective CarSharing companies must guarantee enough vehicles at all times and at all parking stations (Beutler 2004). The concept of 'Open Destination' goes even further to characterize the new free-floating systems, thus expanding the 'One Way'-principle. With 'Open Destination', users do not even have to specify the place of return; it is only required that the car is parked within the operating area of the CarSharing organization. The advantage here is that in case of disturbances in traffic flow (traffic congestion for example), the CarSharing vehicle can be parked and another mode of transportation can be used. One-way trips generally allow users to adopt intermodal transportation and to use different means of transportation when travelling to and from a certain location. This is not possible in station-based CarSharing as the car must be returned to a specific CarSharing station.

*Costs and billing*

CarSharing companies employ combined tariffs that are dependent on time and distance. It is often the case in station-based CarSharing shemes that the vehicle must be booked for a minimum of one hour.

Different CarSharing companies have varying philosophies regarding tariffs and some choose to charge the distance covered and apply time-dependent fees as well. The rates can also be occasionally differentiated among target groups, times of the day and types of cars. This old structure of pricing is being increasingly abandoned though, as it is first off no longer considered appropriate and secondly, more comprehensive, usage-dependent variations of fees are sought (Sakhdari 2006). Meanwhile, a differentiation in pricing models has been evolved, such as packages for large distances and times travelled, differences between day and night hours, and partial half-hour billing (Mobil in Deutschland e.V. 2011). In order to avoid unnecessarily complicated situations for the user and to increase transparency, new developments are moving more in the direction of a purely time-dependent tariff and a more precise, per-minute charge (Schwieger 2004). In many tariffs, the costs of fuel, parking, insurance and depreciation are included. Payment occurs most often via bill or credit card where a list of all trips is provided. As a general rule, the vehicles are insured through the tariff which includes an excess from the user (TCRP 2005).

In summary, two different types of CarSharing exist today: station-based and free-floating CarSharing. However, these two types reflect two very different forms of CarSharing. Developments have shown that traditional, station-based CarSharing schemes adopt elements from free-floating models and vice versa. In addition, hybrid forms of CarSharing include, for example, neighbourhood-based systems where vehicles are equipped with residential parking permits for one particular controlled parking zone.

The following two chapters give a description of the free-floating CarSharing system *DriveNow* and explain how the member and usage structure of *DriveNow* differ in comparison to traditional, station-based CarSharing systems.

## The example of *DriveNow*

Since June 2011, BMW has been providing the free-floating CarSharing system *DriveNow* in cooperation with the car rental company Sixt. This joint venture has approximately 300 BMW and MINI vehicles in Munich: MINI COOPER, MINI CABRIO, MINI CLUBMAN, BMW 1 Series and BMW X1. It is the first CarSharing service with premium vehicles. At the moment, *DriveNow* is also available in Berlin and Düsseldorf and will soon be available in other European cities. Membership with *DriveNow* requires a one-time registration fee of €29.00; other costs are dependent on use.

From an organizational perspective, *DriveNow* is a station-independent, free-floating CarSharing model. This means that there are no fixed pick-up and return stations. Instead, DriveNow has a defined operating area where vehicles can be picked up and dropped off independently. Thus, both 'One Way' and 'Open Destination' are incorporated in the *DriveNow* model. The operating area in Munich covers approximately the entire arca within the Mittlerer Ring (see Figure 11.1); this was established in order to guarantee a distance of no more than 500 metres from any given location to the next vehicle. Vehicles are spread throughout the operating area and are allowed to park in free, on-street parking spaces. The vehicles are equipped with parking vignettes, which have been bought by the operator from the City of Munich and tested in a pilot phase for a period of 48 months. The parking vignette entitles the vehicle to park in all metred on-street parking spaces that are in the parking-regulated zone, with the following exceptions: Innenstadtviertel within the Altstadtring, the Hauptbahnhofviertel and within residential-only parking zones. The advantage of this regulation for the user is simple: the user does not have to buy a parking ticket himself; rather, the parking fees are included in the price per minute. The operator, rather than the user, pays a fixed fee for each separately issued vignette.

Available vehicles can be found on the internet as well as via phone or smartphone application. If necessary, a vehicle can be reserved 15 minutes in advance in order to ensure that it is available upon arrival. If the car is not occupied within these 15 minutes, it is made available for other users. It is also possible to book an available vehicle spontaneously and directly. The booking and user identification can be found via a chip (*DriveNow*-ID) attached to the users driver's licence, which is held against the reading device on the windshield of the vehicle in order to open the vehicle. Once in the vehicle, a built-in smartphone prompts the user to enter their personal identification number and guides them through the booking process as well as offering further information. *DriveNow* users have to pay €0.29 for each minute driven. This price includes the costs for fuel, insurance,

**Figure 11.1 The *DriveNow* business area in Munich**

taxes and parking fees. Furthermore, in case the driver stops during the booked period, €0.10 per minute is charged. In addition, users are offered three-, six- and nine-hour packages, which all include a certain amount of kilometres. All billing occurs via credit card.

## Member and usage structure of *DriveNow* in comparison to station-based CarSharing

The following section provides an overview of the member and usage characteristics of the free-floating CarSharing system *DriveNow* and compares this data to previous knowledge about CarSharing usage patterns. The analysis of *DriveNow* takes around 11,000 active members into account, all of whom have driven at least once in Munich between June 2011 to June 2012. As a basis for analysis, user and booking information as well as the results of an online survey in July 2011 were provided by *DriveNow* (DriveNow 2012). The user and booking data contain various information about more than 170,000 trips made by *DriveNow* users during this period, including User ID; gender; age and date of registration; time and position at the start, intermission and end of booking; distance travelled and vehicle identification number. For additional information on socio-economic characteristics and purpose of use, an online survey with a period of five days was conducted in July 2011, just one month after the launch of the system. The invitation for the survey was sent via newsletter and the scheme's Facebook blog. In total, 204 *DriveNow* users participated, which corresponds to a representative sample of around 5 per cent of all registered users at that time.

The CarSharing member and usage structure has been rather distinct until now. Typical users tend to be those who are ecologically oriented and who often identify themselves as politically left-wing. It can be ascertained that more men choose CarSharing than women, with approximately two thirds of members being male (Huwer 2003, Cairns 2011). From this, it can be assumed that a lower affinity for cars and technology exists amongst the female population and that men tend to try innovations earlier than women. Such assumptions also hold for *DriveNow*: most of the active *DriveNow* members in Munich are male (80 per cent) (DriveNow 2012).

In studies regarding station-based CarSharing schemes there is a predominance of younger people using this kind of mobility service; half of the members are generally between the ages of 25 and 34. The average age of the users tends to fall between 25 and 40. Research has also shown that the younger population (under 25 years of age) and the older population (over 40) are strongly underrepresented (Cairns 2011). These findings can also be recognised among *DriveNow* members: the average age of *DriveNow* members is 34 years while the most frequent age is 30 years.

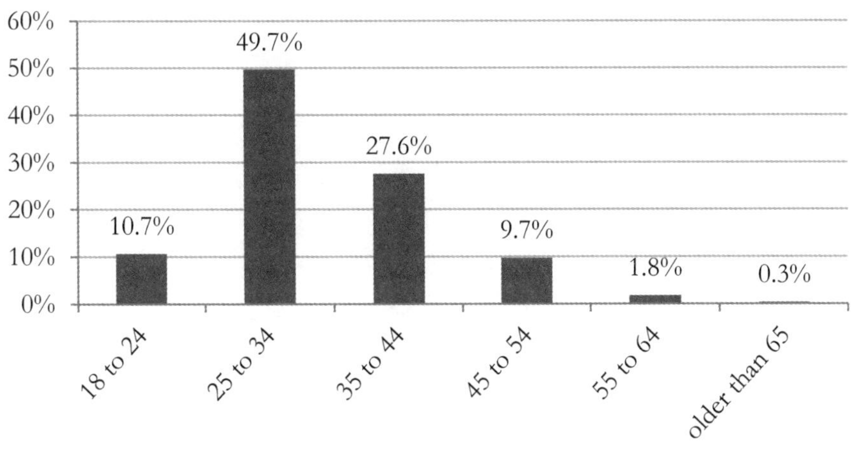


**Figure 11.2 Age distribution of Munich *DriveNow* members (n = 10046)**

Almost half of the active Munich *DriveNow* members are between 25 and 34 years old; the older age group of over 40 is strongly underrepresented. In fact, only 2.1 per cent of all active Munich members are older than 55 years. This might be because older people tend to be afraid of using technology such as smartphones or because they tend to have consolidated mobility routines. Therefore, it might be difficult to satisfy the needs of the older age group, especially when they have to adapt to different types of vehicles. 10.7 per cent of active members are young drivers (under the age of 25), many of which are students or trainees (DriveNow 2012). For this user group, CarSharing provides a possibility to get access to individual transport options and to gain driving experience without owning a car.

In general, CarSharing members tend to live in inner-city areas and respectively in neighbourhoods with high residential densities (Muheim 1998). Here, alternatives to car transport are provided, such as fully developed public transport systems and high-quality infrastructure for non-motorized modes of transport. Also, restrictions on owning a car are higher in dense urban areas: parking costs are high due to the lack of parking spaces in urban areas. Traffic congestion and other restrictions for car drivers, for example the 'Umweltzone' and the prioritization of public and non-motorized transport, additionally limit car use strongly.

Munich is the German city with the highest population density, having an average of 4.355 residents per km² compared to 229/km² in Germany (Statistisches Bundesamt 2012).

**Figure 11.3 Distribution of *DriveNow* members in Munich (n = 10046)**

Therefore it does not seem surprising that 93 per cent of the active *DriveNow* members live within the City of Munich: 77 per cent of the members live within the designated business area and 56 per cent of the *DriveNow* members live within the Controlled Parking Zone (DriveNow 2012). Interestingly, the share of the *DriveNow* business area in comparison to the whole urban territory only stands at 22 per cent and the share of the whole controlled parking zone area amounts to 8 per cent. This result confirms the assumption that CarSharing is more attractive for users who live in dense areas, with a high possibility of traffic and where it is more burdensome to keep an own car.

The civil status of *DriveNow* members was surveyed online in July 2011. Most of them live in households without children, as partners or alone (single). Only around 15 per cent of the *DriveNow* members live with children in their household (DriveNow 2012). Compared to these findings, the civil status of the Munich population is slightly different: fewer people live in a partnership without children in the household (25.7 per cent) while more people are single without children (54 per cent). The distribution of Munich citizens who live in partnership with children (16.9 per cent) or are single with children (3.4 per cent) is almost the same (Landeshauptstadt München 2012).

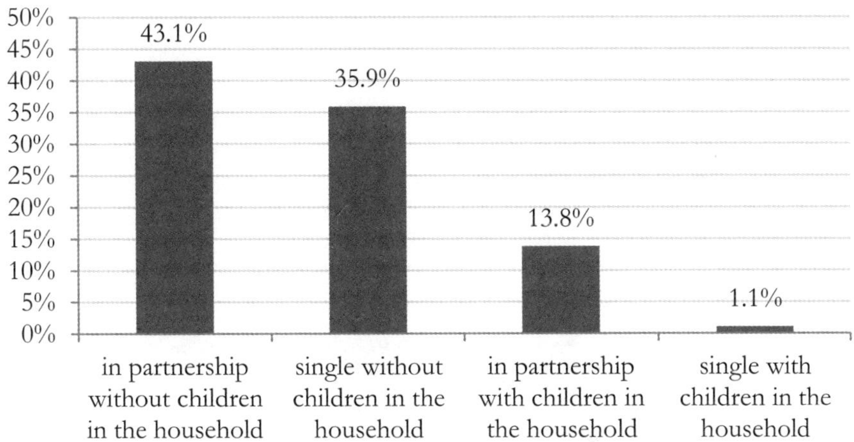


**Figure 11.4 Civil status of Munich *DriveNow* members (n = 171)**

This result confirms the outcome of several CarSharing studies: most of the traditional CarSharing users live in households without children (Muheim 1998). However, it is likely that people who live together with children more often own a car to organize their everyday life.

Furthermore, CarSharing members tend to earn above average incomes and have higher education levels; those with a university degree tend to predominate the traditional CarSharing user group (80–90 per cent of the members).

When comparing Munich *DriveNow* members to the average population, the proportion of *DriveNow* members with a university degree is much higher than the average population; on the other hand, when comparing this measurement to station-based CarSharing members in general, the percentage of members with a university degree in *DriveNow* is significantly lower (DriveNow 2012).

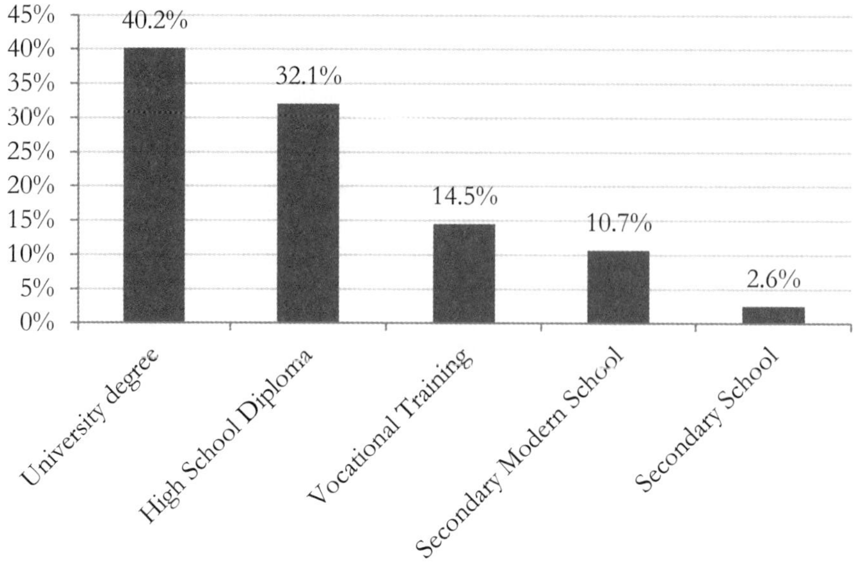


**Figure 11.5 Education level of Munich *DriveNow* members (n = 204)**

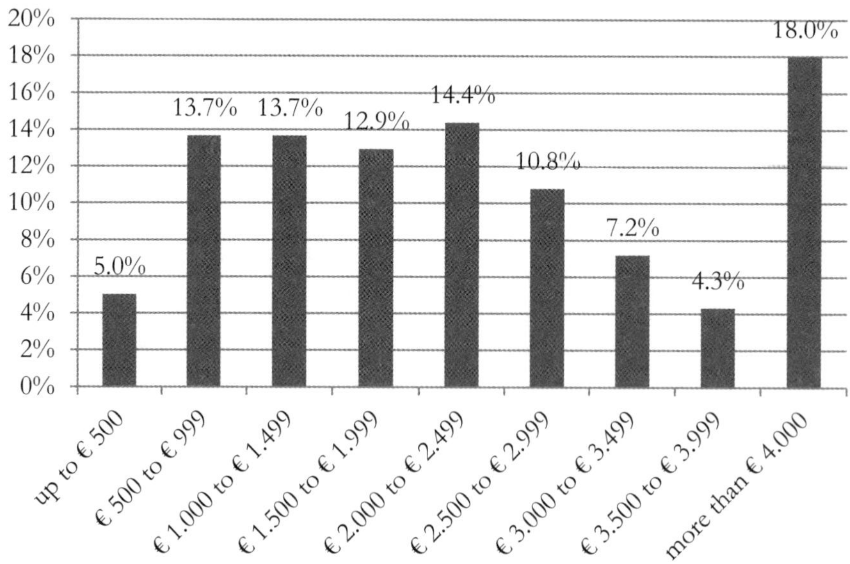


**Figure 11.6 Monthly personal net income of Munich *DriveNow* members (n = 139)**

For station-based CarSharing, it is well-known that members with a low income are strongly under-represented (Loose 2004). In most cases income is related to the level of education. Therefore, it can be assumed that the distribution of income among *DriveNow* members is comparatively high.

Almost one-fifth of the surveyed *DriveNow* members have a personal net income of €4,000 per month. This confirms findings from previous studies on CarSharing, which state that members usually have a higher-than-average income. In Munich, the average monthly net income amounts to €1,963, which means that almost 55 per cent of Munich *DriveNow* members have a higher-than-average personal income (Bayerisches Landesamt für Statistik und Datenverarbeitung 2012).

It is worth noting that CarSharing users tend not to be frequent drivers, which explains why CarSharing is mostly used for leisure-time activities, vacations, shopping and transport (Sakhdari 2006: 58). The *DriveNow* survey also shows that CarSharing is used by a fourth of the members to supplement other modes of transport, which is only possible because of the 'One Way' or 'Open Destination' component: an intermodal link between different modes of transport may be, in fact, only be possible when the chosen mode of transport must not be taken back to the starting point (DriveNow 2012).

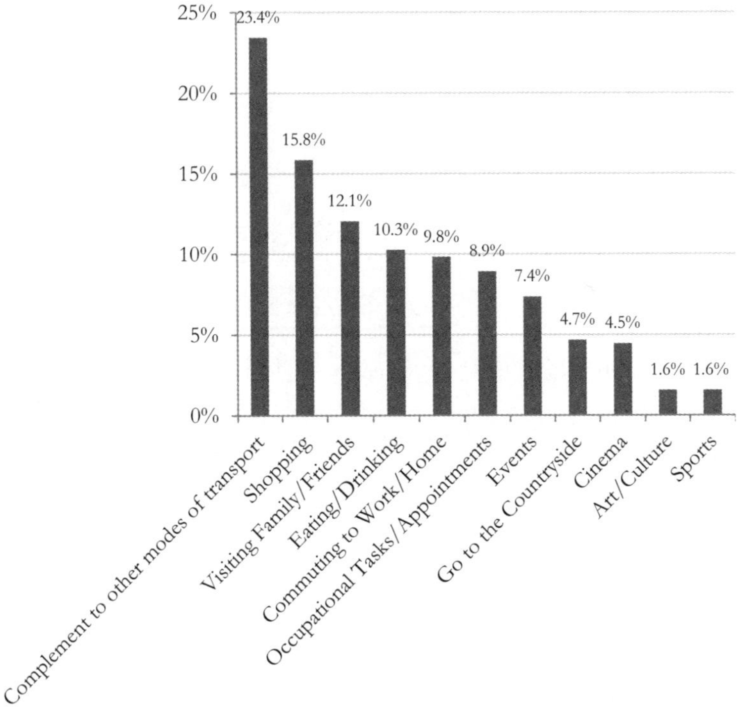


**Figure 11.7 Purpose of *DriveNow* usage in Munich (multiple answers possible; n = 448)**

In addition, users of *DriveNow* specify usage purposes similar to users of classic forms of CarSharing: both state that CarSharing is mainly used for shopping or leisure activities. This is common when users have something that is difficult to carry, somewhere that cannot be served with other means of transportation, or some planned activity that requires a high degree of flexibility. This includes, for example, trips in regions not accessible by public transport or non-motorized modes of transport, or trips that can generally not be achieved without a vehicle. Therefore, for regular trips, members rarely fall back on *DriveNow* vehicles. While 10 per cent of respondents mention using the service to commute to work, in most cases these commutes have been one-way trips, such as shopping, that are made after work (DriveNow 2012).

Studies regarding members of traditional CarSharing schemes show that for other purposes, public transport is usually preferred and CarSharing members most often have a membership or long-term ticket for public transport (Nobis 2006). Also, the *DriveNow* survey in Munich shows that public transport and bikes are preferred as the main means of transportation. Taxi, rental cars and traditional schemes of CarSharing are used barely or very rarely by *DriveNow* members (DriveNow 2012).

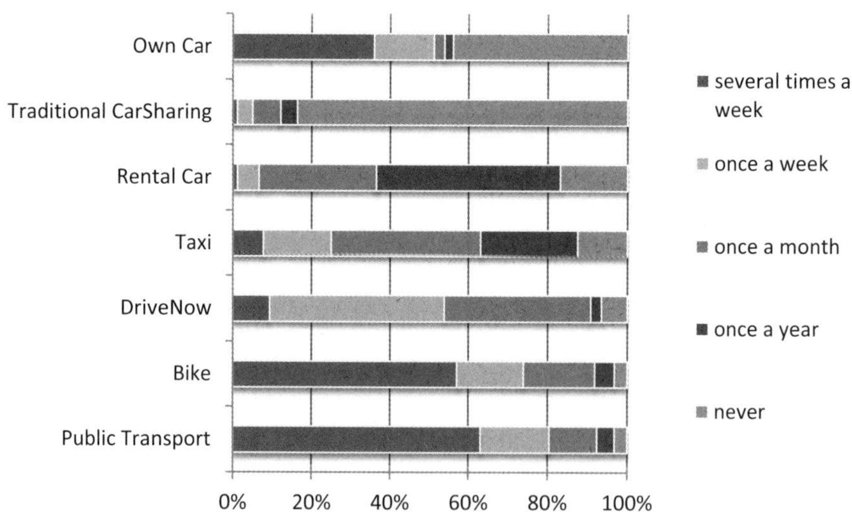


**Figure 11.8 Mode choice of *DriveNow* members (n = 1,288)**

Here, the survey has indicated that the main frequency of members using *DriveNow* ranges from once a week to once a month. In terms of the actual distribution of this frequency though, records found that *DriveNow* members in Munich mostly use a vehicle either several times a month or several times per year. Thus the actual records show a lower frequency than stated in the survey (DriveNow 2012). In this analysis, the duration of user membership was taken into account; the number of trips calculated were dependent on the registration date. Ultimately, results show that *DriveNow* is not the principal mode of transport and will always be used in combination with the other means of transportation offered.

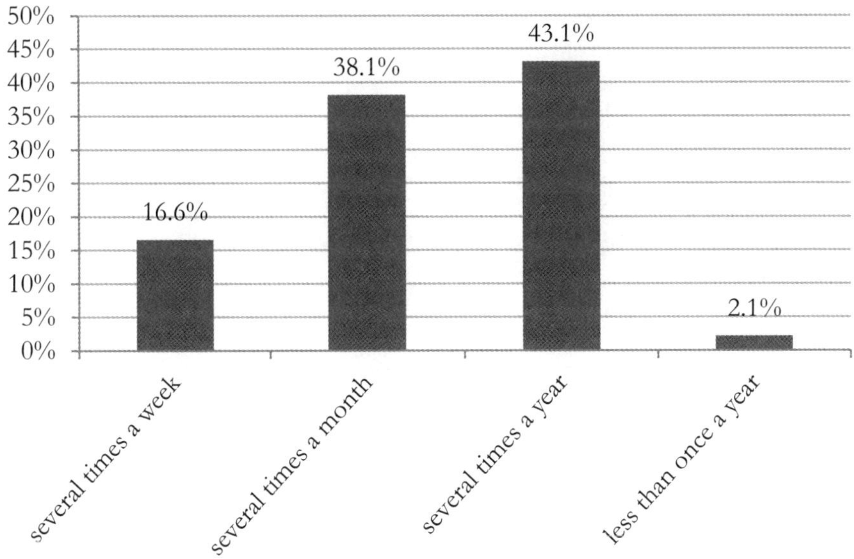


**Figure 11.9 Frequency of *DriveNow* usage (n = 10046)**

These results confirm the outcome of several CarSharing studies: The majority of traditional CarSharing customers tend to use a vehicle one to two times a week (Loose 2004: 3). On average, CarSharing members use a vehicle less than once a month, while less than 4 per cent hire a vehicle on a weekly basis or more frequently (Cairns 2011). It is hardly possible to compare the distance travelled and the time of use for a free-floating system like *DriveNow* with station-based schemes of CarSharing though. This is a result of a free-floating system's systemic nature, which can be seen in pricing differences, but is especially due to the 'One Way' opportunity, where it can be assumed that distance travelled and usage are fundamentally different from those of traditional CarSharing. At first, even if a *DriveNow* vehicle would be used for one round trip, it could be a different vehicle.

Measuring one-way trips is extremely difficult, as assumptions have to be taken, which can sometimes affect the outcome greatly. Looking at the distribution of distance travelled, the majority of trips in *DriveNow* are between four to seven kilometres and almost 50 per cent of the trips are longer than seven kilometres, which cannot be considered a long distance for motorized vehicles (DriveNow 2012).

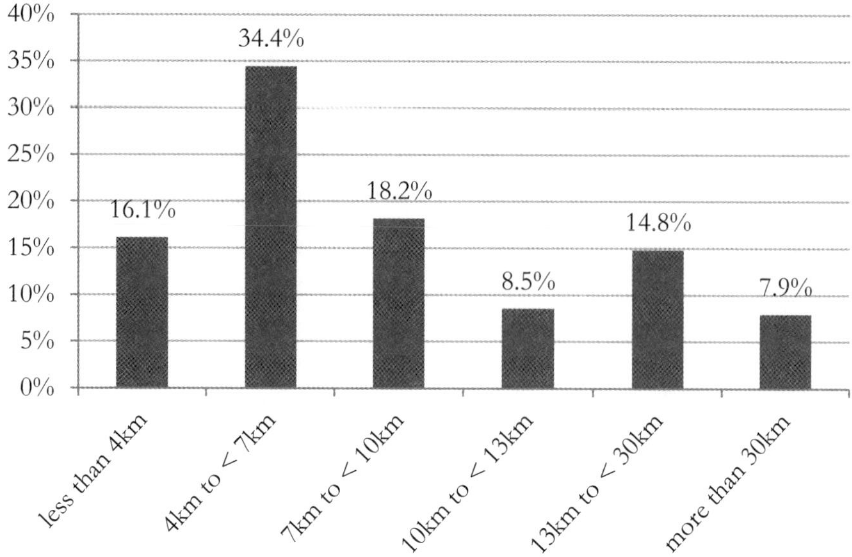


**Figure 11.10 Trip distances of Munich *DriveNow* members (n = 10046)**

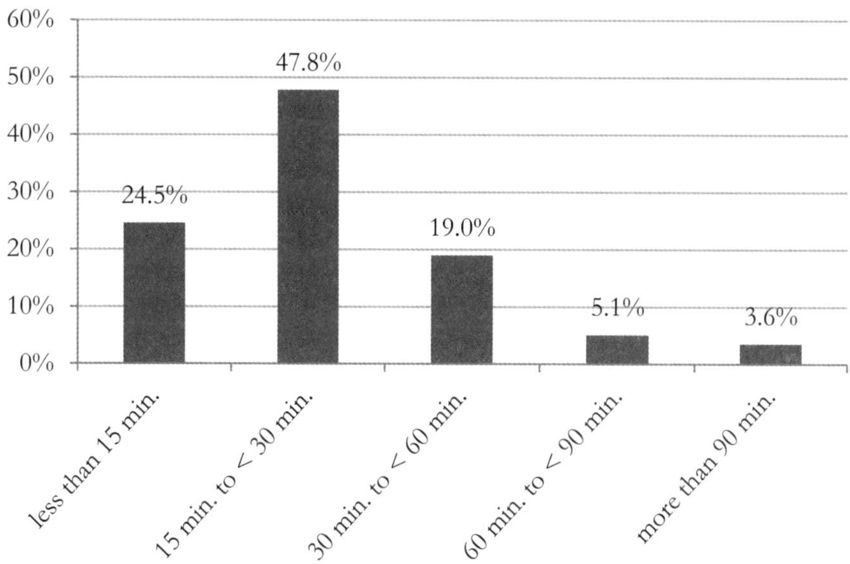


**Figure 11.11 Duration of *DriveNow* usage (n = 10046)**

A trip, on average, is about 13 km long and has a high standard deviation of 21. The distance travelled should not necessarily be related to the booking period though, since this measured distance is made up of both driving time and parking time. The predominant part of the bookings take under an hour and less than 10 per cent of bookings last for more than two hours (DriveNow 2012).

On average, the booking period is 48 minutes, consisting of 29 minutes of driving time and 19 minutes of parking time. Looking at how the booking period is composed in different categories, the proportion of parking increases for bookings that have a longer duration (DriveNow 2012).

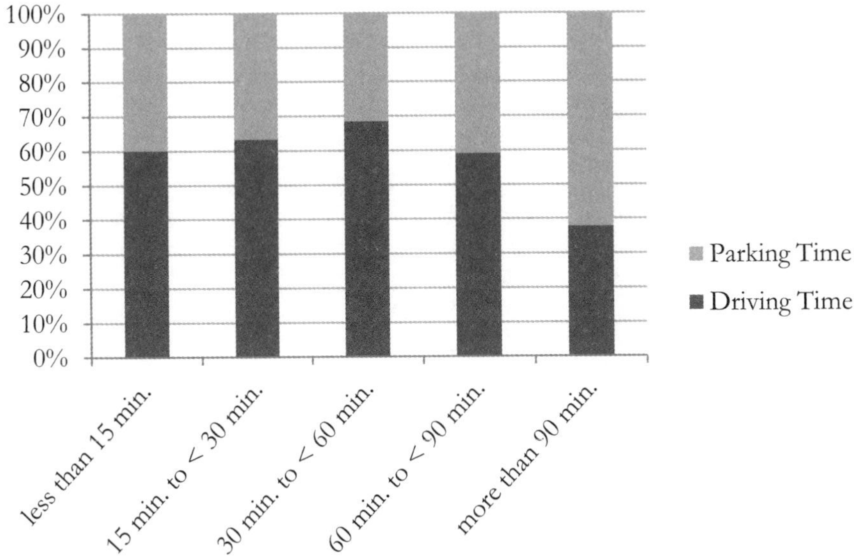


**Figure 11.12 Distribution of *DriveNow* travel time (n = 10046)**

These are trips where a destination lies outside of the business area and thus where the vehicle is parked during the scheduled activity because the vehicle cannot be returned in this specific location outside.

**Prospects for CarSharing schemes**

CarSharing users belong to a homogeneous milieu that has a certain distinctive social profile. Over time, this profile has proven itself to be quite stable, but it is worth noting that *DriveNow* members do not belong to the previously rigid environment of CarSharing users. However, CarSharing as a mobility service is an attractive offer for a much broader audience due to low entry barriers and minimal, usage-based costs. On the other hand, it is clear that completely new target groups, which have until recently played a smaller role in traditional CarSharing models, can be addressed by a free-floating CarSharing offer such as *DriveNow*. Here, it seems likely that this development can be justified to a large extent by the technical developments of the system and the resulting comfort for the user.

In addition to the technologies for vehicle access and the opportunities offered by computer-based accounting and billing processes, it should be mentioned at this point that the proliferation of internet-enabled customer electronic devices has especially made spontaneous car use outside the home possible and much easier.

It remains to be seen whether free-floating CarSharing systems show a similar traffic impact as documented for classic CarSharing models in many studies, especially in terms of motorization and traffic performance of the participants. It should be noted, however, that the effects of CarSharing on the whole transport system are methodologically difficult to measure and a direct connection between CarSharing membership and positive environmental effects cannot be detected conclusively due to dynamic systemic reactions.

Nevertheless, it can be assumed that among members of advanced free-floating CarSharing models, pragmatic inter- and multi-modal transport use is observed in similarity to traditional CarSharing and is caused by the cost transparency; CarSharing is used for specific needs which cannot be covered with public transport or non-motorized vehicles.

The use of electric vehicles for CarSharing seems to be particularly interesting and therefore, mutual leverage is expected to result: the use of electric vehicles could make CarSharing more attractive, especially for people who want to test an electric vehicle without investing in its currently higher costs, when compared to a conventional combustion engine vehicle. On the other hand, CarSharing could represent an important area of the use for electric vehicles, which so far only accounts for a small number of registered vehicles. Nevertheless, it remains to be seen whether and how the use of electric vehicles in free-floating CarSharing models can be organized, especially since the charging infrastructure is still station-based.

## References

BASt – Bundesanstalt für Straßenwesen 2004. *Bestandsaufnahme und Möglichkeiten der Weiterentwicklung von Car-Sharing*. Bergisch Gladbach.

Bayerisches Landesamt für Statistik und Datenverarbeitung 2012. *GENESIS-Online Datenbank* [Online]. Available at: https://www.statistikdaten.bayern.de/genesis/online/data?operation=abruftabelleAbrufen&selectionname=82411-001z&levelindex=1&levelid=1351527489698&index=2 [accessed: 25 May 2012].

BCS – Bundesverband CarSharing e.V. 2011. *Jahresbericht 2010*. Berlin.

BCS – Bundesverband CarSharing e.V. 2012. *Marktanteil CarSharing-Anbieter in Deutschland – Nutzungssysteme nach CarSharing-Teilnehmern* [Online]. Available at: http://www.carsharing.de/images/stories/pdf_dateien/grafik_marktaufteilung_nutzer_nach_system_2012.pdf [accessed: 25 May 2012].

Beutler, F. 2004. *Intermodalität, Multimodalität und Urbanibility – Vision für einen nachhaltigen Stadtverkehr* (Discussion Paper SP III 2004-107). Berlin: Wissenschaftszentrum Berlin für Sozialforschung.

BMW Group 2011. *BMW Group und Sixt AG gründen Joint Venture DriveNow für Premium Car Sharing – Innovatives Mobilitätskonzept für Innenstädte* (Presse Information). Munich.

Cairns, S. 2011. *Accessing Cars: Different Ownership and Use Choices*. London: RAC Foundation.

Canzler, W. 2010. Mobilitätskonzepte der Zukunft und Elektromobilität, in *Elektromobilität. Potenziale und wissenschaftlich-technische Herausforderungen*, edited by R. Hüttl, B. Pischetsrieder and D. Spath. Berlin: oekom.

DriveNow – DriveNow GmbH and Co. KG 2012. *Member and Usage Data Munich. June 2011 to June 2012*. Munich.

Fliegner, S. 2002. *Car Sharing als Alternative? Mobilitätsstilbasierte Potenziale zur Autoabschaffung*. Mannheim: MetaGIS.

Franke, S. 2001. *Car-Sharing – vom Ökoprojekt zur Dienstleistung*. Berlin: Edition Sigma.

Harms, S. 2003. *Besitzen oder Teilen: Sozialwissenschaftliche Analyse des Car Sharings*. Zürich: Rüegger.

Huwer, U. 2003. *Kombinierte Mobilität gestalten: Die Schnittstelle ÖPNV – CarSharing*. Kaiserslautern. Ph.D. Thesis. http://nbn-resolving.de/urn:nbn:de:bsz:386-kluedo-15370 [accessed: 22 July 2012].

ITS – Institute of Transportation Studies UC Davies 2008. *Worldwide Carsharing Growth: Am International Comparison* [Online]. Available at: http://escholarship.org/uc/item/1139r2m5 [accessed: 13 March 2012].

Johnsen, D. 2007. Carsharing – Das Dilemma zwischen Privatwirtschaft und gesellschaftlichem Auftrag, in *Möglichkeitsräume innovativer Mobilitätsdienstleistungen als ergänzende Funktion des ÖV*, edited by Deutsches Zentrum für Luft- und Raumfahrt DLR. Berlin. http://elib.dlr.de/51401/ [accessed: 22 July 2012].

Koss, R. 2002. *Car-Sharing als Beitrag zur Lösung der verkehrs- und umweltpolitischen Krise?* Berlin: VWF.

Landeshauptstadt München 2012. *Private Haushalte* [Online]. Available at: http://www.muenchen.de/rathaus/Stadtinfos/Statistik/Bev-lkerung/Haushalte.html [accessed: 11 May 2012].

Loose, W. 2004. CarSharing – Potenziale für weniger Autoverkehr, in *Handbuch der kommunalen Verkehrsplanung. Für Praxis in Stadt und Region*, edited by T. Bracher, H. Holzapfel, M. Lehmbrock, M. Haag, F. Kiepe and U. Reutter. Berlin: VDE.

Millard-Ball, A., et al. 2005. *Car-Sharing: Where and How it Succeeds*. Washington D.C.: Transportation Research Board.

Mobil in Deutschland e.V. 2011. *Mobil in Deutschlands großer Carsharing Atlas 2011* [Online]. Available at: http://www.mobil.org/_news/mobil_in_deutschlands_groer_carsharing _atlas_2011 [accessed: 11 May 2012].

Muheim, P. 1998. *CarSharing – der Schlüssel zur kombinierten Mobilität*. Luzern: Eidgen.

Nobis, C. 2006. *Car Sharing as a Key Contribution to Multimodal and Sustainable Mobility Behavior – the Situation of Car Sharing in Germany.* Berlin. http://elib.dlr.de/22239/ [accessed: 22 July 2012].

Pesch, S. 1996. *Car-Sharing als Element einer Lean Mobility im PKW-Verkehr – Entlastungspotentiale, gesamtwirtschaftliche Bewertung und Durchsetzungsstrategien.* Düsseldorf: Fischer.

Petersen, M. 1995. Ökonomische Analyse des Car-Sharing. Wiesbaden: Gabler.

Rösch, K. 2001. *Car-Sharing.* Ph.D. Thesis. Nürnberg.

Sakhdari, F. 2006. *Vermarktung von CarSharing-Konzepten.* Ph.D. Thesis. Berlin.

Schwieger, B. 2004. *International Developments towards a 'Second Generation' Car-Sharing.* London: Writersworld.

Shaheen, S. 1999. *Dynamics in Behavioral Adaption to a Transportation Innovation: A Case Study of carlink – A Smart Carsharing System.* Davis. http://www.escholarship.org/uc/item/87n6958h [accessed: 22 July 2012].

Shaheen, S. A. and Cohen, A. P. 2007. Growth in Worldwide Carsharing: An International Comparison. *Transportation Research Record 1992*, 81–9.

Statistisches Bundesamt 2012. *Regionalatlas der Statistischen Ämter des Bundes und der Länder* [Online]. Available at: http://ims.destatis.de/indikatoren/Default.aspx?https=1 [accessed: 12 June 2012].

Steding, D., Herrmann, A. and Lange, M. 2004. *Carsharing – sozialinnovativ und kulturell selektiv? Möglichkeiten und Grenzen einer nachhaltigen Mobilität* (ZUFO Berichte, 3). Münster: Universität Münster Zentrum für Umweltforschung.

TCRP – Transit Cooperative Research Program 2005. *Car-Sharing: Where and How it Succeeds* (TCRP Report 108). Washington, D.C.: Transportation Research Board.

Wilke, G. 2009. Ressourcenschonung durch Car-Sharing – Aussichten veränderlich, in *Umwälzung der Erde: Konflikte um Ressourcen*, edited by G. Altner. Stuttgart: Hirzel.

# Chapter 12
# The Social Sustainability of Urban Transport: A Contemporary Reflection on Social Inclusion and Social Exclusion in the Context of Urban Disaster

Margaret Grieco

## Introduction – planning and scale: the promotion of sustainable order

Evidence shows that natural disasters are happening at an accelerated pattern and pace[1] (Freeman 2000) and that current land use and infrastructure planning have been in denial about the implications of the patterning of such natural disasters for the scale, locations and activities of human habitation. Discussions of the social sustainability of urban transport and of urban infrastructure have begun to engage with issues of social equity and social exclusion under routine conditions of urban functioning (Grieco 2007) and to an extent have appreciated that social inequalities are heightened in situations of natural disaster crises. Nevertheless, the appreciation of vulnerability had only been focused on and confined primarily to developing countries until the flooding of New Orleans.

Within the context of the social sustainability of urban transport, it is clear that mega cities and dense settled geographical areas have to extend a greater consideration to the pattern of geographical risk factors associated with their locations. Further, they need to develop strategies for evacuation, for the potential de-densification of existing cities and for the re-establishment of their populations and communities elsewhere. The tradition and legacy of locating human habitation on coastal plains and along river banks is readily understandable as an outcome of the economics of past transport and communications structure along with the ease of infrastructure development (especially in regards to waterside property development, which has enjoyed a prime position in contemporary real estate regeneration and leisure location dynamics). The danger of continuing with this tradition, given the pattern of the recent sequence of natural disasters, is clear.

1 Institution of Civil Engineers. *Vulnerability To Natural Hazards – Policy.* Available at: http://www.ice.org.uk/patoolkit/Policy/Sustainable-infrastructure#card_8 [accessed: 18 April 2012].

Similarly, the attractiveness of the large scale as a planning principle has had a long hold on the development of human habitation and the promotion of the large scale in terms of population densities has had its consequence on the scale of buildings, transportation vehicles and other aspects of urban infrastructure. High buildings, congested cities and the comprehensive covering of cities' natural surfaces with cement can accentuate physical destruction through earthquakes, impede the speed of escape from natural disasters and contribute to the impact of flooding by impeding natural drainage potentials. The physical scale of edifices and large items of infrastructure can play its own part in amplifying, accentuating and intensifying the impact of natural disaster.

At the same time, as an accelerating and accentuating pattern of natural disasters occurs, new information communication technologies permit for new patterns of greater distribution of populations within routine circumstances. This can afford technologies for reconnection, repair and revitalization under crises conditions. Planning for reconnection, repair and revitalization after a disaster become a central requirement of the social sustainability of contemporary transport systems and structures: utilizing new technologies to their full value to reduce risk through reducing settlement densities and harnessing them in crisis to enable reconnection requires innovative thinking and action.

To discuss sustainable mobility adequately, we must discuss the sustainability of settlement patterns and the sustainability of infrastructure patterns in the context of the now known grand scale of natural disasters. We must begin to appreciate the different and differing time horizons available for alerting populations – most particularly the most vulnerable segments of communities – to the impending disasters and to get smarter about developing flexible modes of social-technical organization which enable the reconnection and re-establishment of impacted communities after a disaster. Following on from an earlier discussion of transport under crisis conditions (Grieco and Hine 2008), this chapter extends this discourse on the promotion of sustainable order. It uses a case study approach in the highlighting of key issues and lessons learnt from contemporary disasters. It raises the issue of the im/mobilities of place and infrastructure in addition to the im/mobilities of people.

## Flooding, after the deluge: memory and management

> Dr Foster went to Gloucester in a shower of rain
> He stepped in a puddle right up to his middle
> And never went there again.
> English folk rhyme (Opie and Opie 1997).

In July 2007, Gloucester in England experienced its worst floods on record (BBC Gloucestershire 2009a) though the English folk rhyme reproduced above suggests that there may be a folk memory of previous flooding of locally significant proportions ('puddle' probably being a transformation of the word 'piddle' meaning a stream).

> In hours the relentless rain had left hundreds of homes and businesses under water, and thousands of drivers and travellers were stranded as road and rail routes were inundated by fast-rising flood waters.
>
> But that was just the start. Soon the taps ran dry for half the county's population as a water treatment station on the outskirts of Tewkesbury was overwhelmed by the floods.
>
> And emergency services faced a desperate battle to save the county's power supplies as the Walham electricity sub-station near Gloucester also threatened to fall victim to the rising flood-waters (BBC Gloucestershire 2009b).

This short report taken from the BBC gives us an indication of the scale of the flooding event – it generated Britain's biggest peace time emergency within this region. The low lying location of fundamental infrastructure such as electricity sub-stations and water treatment plants heightened the scale of the disaster beyond the swamping of transport routes.

> Mythe water plant at Tewkesbury flooded leaving 150,000 houses without water for two weeks. The Army and emergency services fought a desperate battle to protect the Walham electricity substation in Gloucester. Only inches stood in the way of hundreds of thousands of homes losing power (Gloucestershire 2009).

The limits of intensive road-based developments through settlements within and along a flood plain were revealed. Dr Foster's message and warning had clearly not been heard by contemporary society: after the flood and after the deluge, policy should be concerned with both the location of settlements and the location and the design of infrastructure such as electricity sub-stations and water treatment plants. Planning for routine circumstances clearly did not meet the requirements of disaster but the treatment of the flooding has largely been framed in terms of a freak event (Gloucester City Council 2011) and the budgets allocated to fundamentally restructuring the settlement pattern, its mobility arrangements and its servicing remain correspondingly minimal. The focus has been on achieving the return to normality: the discussion of replanning the location of infrastructure to avoid future floods or redesigning infrastructure to withstand the impact of future floods has not received the attention that the worst peace time emergency should have provoked.

Dr Foster's message, much amplified by nature and by man's interaction with nature, is still not heard in the United Kingdom. In the Netherlands, with its low lying geography and its correspondingly constant battle to prevent flooding, a different strategic approach to dealing with flooding is emerging: the doctoral work of Rutger de Graaf on floating urbanization gives a taste of this new vision (De Graaf 2009) with his discussion of the current luxury development of 'floating houses':

> Floating urbanisation has the potential to contribute to reduce vulnerability of delta areas. During floods, floating constructions increase the coping capacity of an urban area. No damage to the construction will occur because floating houses will adapt to the rising water level. In addition, they may serve as emergency shelter during flooding. Because floating houses can be relocated, they are also flexible and reversible, which is a benefit to deal with uncertain future developments such as climate change. This contributes to strengthening of adaptive capacity. Floating constructions are in particular interesting for new urban areas. The potential of floating urbanisation to reduce vulnerability of existing urban areas remains limited. It is not feasible to change current urban areas into floating urban areas. The majority of future urban areas are already there. For these kinds of areas, other solutions are necessary to reduce flood vulnerability.

Floating houses in the Netherlands give a suggestion as to what might be accomplished on the domestic accommodation front already today, even though the policy has not yet developed to move this new technical capability forward at any significant scale. The development of floating domestic architectures also raises the issue of what might be the case in the future with the design of larger scale infrastructure services.

New Orleans shared the challenge of its flood prone geography with the Netherlands and developed fixed infrastructures to meet the challenge. Nevertheless, the breaching of these infrastructures by a dramatic – but predictable – weather event (Hurricane Katrina in 2005) saw a major catastrophe. The failure to adjust settlement planning to reduce vulnerability to floods was compounded in New Orleans by the failure of transport planning to ensure that adequate provision was made for transit dependent inhabitants to enjoy an orderly evacuation from the city when the flooding alert was raised. Todd Litman (2006) provides a chilling assessment of the various level of planning failures exhibited around Hurricane Katrina and the flooding of New Orleans – the 'planning' documents indicate that the vulnerability of the transit dependent residents was known and the level of their immobility in the context of predictable disaster was left largely unaddressed.

Elsewhere (Grieco and Hine 2008), we have discussed the issue of the link between stranded mobility as exhibited in New Orleans in both routine and crisis conditions and in the social capital structures of the black population of New Orleans in the context of the limited ability of Louisiana as state and hinterland to absorb the displaced population of New Orleans. Substantial components of the black population of New Orleans neither had people nor place to escape to. We also drew attention to the challenges that would be present in the reconstruction of the destroyed black neighbourhoods of New Orleans and suggested new technology paths of virtual neighbourhood reconstruction tools that could be used to reconnect and redevelop communities as communities. 'Place' can be reconstructed as 'virtual place' as part of a process of redeveloping and re-establishing material and historic place: memory can be harnessed as part of the management of reconstruction after the flood and after the deluge.

> In line with New Orleans history in the registration of the black vote and the desegregating of schools, the electronic registration of neighbourhoods could be used to ensure that presently displaced voters continue to participate in New Orleans politics. Although, the indications are that the city is not yet ready for the return of all of its old inhabitants, electronic neighbourhoods would provide a basis for future planning which retained the social character of New Orleans prior to the disaster. An electronic Lower Ninth could participate in the reconstruction or relocation of the Lower Ninth as an entity whichever situation is the final outcome of New Orleans' new planning horizons.
>
> Virtual neighbourhoods will permit virtual travel to New Orleans for those presently located elsewhere. This simple approach does not seem to have been adopted but it is a relatively simple extension of the tracking technologies now planned for use in any future disaster. Virtual mobility may be an important element to combat physical social exclusion (Grieco and Hine 2008).

New information technologies clearly have a role to play in enabling reconnection and reducing the impact of the social capital loss that occurs with the displacing and scattering of the membership of any community. The Lower Ninth Ward of New Orleans half a decade later still remains an obvious candidate for such intelligent policy making given the obvious failure of conventional planning processes more than half a decade after the disaster:

> Five years after Katrina, the Lower Ninth remains a world of hurt. Predominantly African-American and working class, it was the district hardest hit by a storm whose biblical floods, it's now agreed, resulted largely from negligence on the part of the U.S. Army Corps of Engineers, the agency that built and was supposed to maintain the surrounding levees. And yet, post-Katrina, the once thriving community is also the hardest hit by seeming government indifference. Only a fifth of the Lower Ninth's 20,000 residents have returned to live since 2005, in no small part because of inadequate reconstruction funding compared to aid that homeowners in other New Orleans neighborhoods have received, and because of the slow pace of long-promised infrastructure and other community development projects (Padgett 2010).

Flood Street in New Orleans' lower ninth ward has received much press attention: the fit between street name and the experience of the Katrina deluge created an opportunity for memorialization that was ironically celebrated by outsiders. Memorialization is a well-established feature of the human engagement with tragedy and loss: the establishment of monuments and cultural markers around significant individual and collective losses and indeed victories is a universal part of the human landscape. The digital age provides for new electronic forms of memorialization as part of the global landscape. Memorialization in New Orleans provided an online stock of photographs of housing design and flood experience that is worth reflecting on here.[2]

Memorialization is important both for reconnection to community and place and for human recovery from disaster – it is a core element of sustainability and it is also important for creating historic and folk-loric or folk memoric alert systems. Nevertheless, it can be used for more than that as we rethink now how we should live and how we should plan to live. The images of flood damaged housing in low income New Orleans raise the issue of whether housing design could routinely be better formulated to meet the challenge of predictable but not date determinable flooding waters. The Dutch discussion of floating urbanization provides us with one path that ought to be receiving more mainstream consideration: the development of homes above flood water levels. Developing homes above flood water levels can mean the geographical relocation of whole communities but it can also mean raising the level of homes in their current neighbourhood location. With images of the flooded wooden homes of New Orleans in our minds let us turn to the recent floods in Queensland, Australia.

Australia's floods in early 2011, most particularly those occurring in Queensland, provided shocking television images. The sight of the restaurants of the Brisbane waterfront ripped out of their groundings and swirling off haphazardly down the Brisbane River provided a graphic new image of what had previously been prestige waterfront developments. But there was a much quieter story that received less press coverage and attention that we should pay attention to here: the design of the traditional Queenslander house. For those familiar with Queensland, the traditional Queenslander house standing grand on its stilts clearly had benefits for the natural circulation of air in hot and humid conditions. But in addition, this type of house was also perfectly designed to cope with flash flooding. Building residences with an inhabited street level takes away this traditional form of protection against tropical weather.

Additionally, there was some local reflection on this, but not enough, it might be argued, to stimulate the global architectural consciousness of what designs are appropriate if floodplain settlements continue to be built.

> One of the great ironies of the situation is that the famous timber 'Queenslander' houses that were elevated on stumps or long stilts up to 3–4m above the ground are disappearing from Brisbane, replaced by standard houses built at ground level on concrete slabs. Flood waters could rise several meters below the 'Queenslander' before entering the house. The living areas were upstairs and the areas under the house was used for storage and a garage. Has the demise of the 'Queenslander' increased the flood damage to houses in Brisbane? Many people believe that the Brisbane houses were built on stilts (stumps) <u>solely</u> for ventilation, so that air

2 See http://www.asergeev.com/pictures/archives/2006/534/jpeg/20.jpg [accessed: 25 October 2012].

> could circulate under them in the hot summer months. While this is true the other major reason for the elevation was **flooding resistance**, both flash flooding due to heavy summer storms, and riverine flooding (Hub Pages 2012).

It can be argued here that the planning establishment lost sight of the communal memory of the natural experience embodied in the original design of the Queenslander. And furthermore, the opportunity has never been taken to reflect on the transferability of such a design for places that could have benefited from it, for which on the image evidence available New Orleans certainly was a candidate.

It could also reasonably be argued that with adaptation to prevent heat loss, the flood plains of Britain are also candidates. Living on stilts to avoid flood waters, and indeed to achieve a defensible living, has a long tradition and has some interesting modern forms (Façade 2011). Stilts can also provide storage facilities for small boats and is necessary to cope with mobility in conditions of flash and riverine flooding. The role of the 'tinnies' – small domestic scaled boats – in achieving mobility during the Queensland floods has some reportage markers (Dibben 2011) and many visual markers. Ensuring the ready availability of water based transport at the domestic level becomes a critical element of social sustainability in a flash flooding context: designing in the storage for such transport has been far from a routine consideration to date but a rethink of urban mobility to include such consideration is certainly due.

Developing the planning consciousness of the history of place and the natural challenges that place has experienced and embraced should be made easier by the new information technology tools which are now available. Despite this, the organization of these tools into robust community planning systems is still a long way off as the case studies presented here demonstrate. 'Dr Foster went to Gloucester' memorializes experience for us, losing our memories of why traditions took shape and form is an unnecessary risk in contemporary society with our knowledge-banking and experience-matching technical capabilities.

## The day the earth shook: earthquakes, architecture's scale and the decision making around construction

> Engineers have a saying: 'Earthquakes don't kill people, buildings do.' Destructive earthquakes have taught New Zealanders hard lessons in designing safe buildings (McSaveney 2009).

The earthquake experiences of Pakistan and New Zealand differ in many respects, but perhaps the two most striking differences are in the scale of the architecture that was involved and the scale of the number of deaths that resulted. Whilst the scale of the architecture involved in the Pakistan earthquake of 2005 (Grieco and Hine 2008) was typically smaller (though more of it was involved as the shake affected a greater region) than the New Zealand quake of 2011, the routine failure to 'earthquake proof' lesser scale domestic building stock in Pakistan (Naeem 2005) generated a very large death toll. Its estimates vary, but over 75,000 deaths and approximately 100,000 injuries appears to be a reliable lower limit, whereas the death toll in the larger scaled buildings of New Zealand was 180.

The loss of life in New Zealand was due to a failure in the engineering knowledge available; the loss of life in Pakistan was due to the failure to apply the engineering knowledge available.[3] The domestic building stock in Pakistan had not been fitted with the appropriate earthquake proofing technologies available. Had the engineering knowledge been applied, the scale of construction collapse would have been minimal in Pakistan but the scale of construction given the failure of engineering knowledge certainly contributed to the death toll in New Zealand – a death toll which was nevertheless less than the previous large scale Napier earthquake disaster of 1931.

Christchurch, where the New Zealand quake struck, had had previous earthquake experience (McSaveney 2009) and contemporary buildings had been designed according to the principles of seismic resistance or had been adjusted through the principle of seismic retrofitting that buildings in Pakistan lacked. Despite this earthquake proofing design, the large contemporary buildings in Christchurch nevertheless collapsed.

The scale of architectural collapse in New Zealand contrasts with the planning literature preceding it, which has a tone of confidence as to the effectiveness of the practice of seismic resistance planning and retrofitting.

> Modern buildings can be designed to be safe under extreme earthquake attack with collapse being avoided. Current earthquake design practices achieve this by dictating the post-elastic response of the building, locating and detailing zones within the structure where high post-elastic deformations are acceptable, rigorously detailing these zones so they can dependably resist the imposed actions while other, less desirable, post-elastic mechanisms are suppressed. Simple procedures for achieving these objectives are included in the paper and their implications on material property (particularly their over-strength ratios) are outlined. Earthquake Loads & Earthquake Resistant Design of Buildings (King 1997).

This assessment contrasts strongly with the commentary that follows the Christchurch quake.

> The powerful 6.3 magnitude shallow earthquake that hit New Zealand last week struck less than 6 miles from Christchurch's historic downtown district, and only about 3 miles below the surface. Engineers were not surprised by the damage

3 After the earthquake and its devastating toll, the Aga Khan foundation has made resources available to allow redevelopment to occur within the context of earthquake preparedness technologies (Aga Khan Development Network 2006 and 2008).

> and tumbling of the older structures, since the epicenter of the quake was so close to the city, but the amount of damage to the more contemporary structures which had seismic retrofitting, was not anticipated.
>
> With so many buildings in the United States using very similar, if not the same building standards as New Zealand, it is important for contractors to study the effects of this quake, before designing any new structures … .
>
> Nevertheless, there is already evidence that some of the buildings that had received seismic retrofitting failed. Amir S.J. Gilani, a structural engineer, said that Christchurch Cathedral, "which crumpled this week, had been retrofitted in the late 1980s and seemed to have performed well [during] last September[s earthquake]." Gilani continued to say that, "the retrofitting project had been used to showcase … advances in this field, however, it certainly experienced much larger accelerations in the 2011 earthquake for which it was obviously not prepared for".[4]

The New Zealand experience has caused concern about the engineering knowledge that informed the planning of construction in that location and more widely, not least within the United States. At the very least, the experience of New Zealand must create concern about issues of scale in respect of building size and density until the requisite engineering knowledge is 're-gained'.

In the context of disasters, mobility becomes critical. Routine mobility deficiencies – or transport exclusion – become accentuated and as we have already seen there may be a very real pressure in a range of disaster circumstances to shift mode from road based to water based mobility. Maintaining dual mode capacities may become a more regular feature of modern life given the new disaster patterning that is being evidenced and witnessed. But even where the mode remains the same, the routine patternings of routes and journeys will be placed under pressure and new arrangements will need to be made. Barriers to mobility prevent help from getting in and the vulnerable from getting out of further disaster-prone and dangerous conditions that follow in the quake of the initial disaster. In the case of Pakistan, rural areas with a heavy population and poor connections to infrastructure can be viewed as experiencing 'stranded mobility'. Remoteness can be defined as the lack of access to infrastructure which connects with facilities; it is not necessarily a register of low population densities

The difficulties experienced in reaching the trapped, retrieving the dead and treating and transporting the injured provide an indication of the low level of facilities routinely enjoyed within the region of North West Frontier Province and Kashmir, where the earthquake hit. Issues of accessibility can accentuate and amplify difficult conditions resulting in corresponding increases in the death toll. The disaster renders visible the poor routine quality of local access to healthcare

4 See http://www.constructionlawsignal.com/by-subject/design-and-technology/new-zealands-quake-topples-contemporary-buildings---even-those-with-seismic-retrofitting/ [accessed: 26 October 2012].

within the affected region: it also reveals the levels of poverty and isolation from other social resources. Poor access in the region and the use of horse-drawn transport clearly impact on the speed with which aid and resources can reach the region.

The disproportionate vulnerability of the aged, sick and less mobile to death in crisis conditions is fast becoming a remarked upon aspect of the pattern of coping with disasters. Despite this, the mobility planning necessary to meet and offset these vulnerabilities is largely absent. Japan is becoming an important venue for the contemporary discussion of these issues. For the moment, we reflect that one major difference between the earthquake experience of New Zealand and Pakistan was that in Pakistan the weaknesses of routine mobility provisions were heightened by the disaster and became life threatening with high levels of stranded mobility whereas this was not a significant element of the New Zealand experience.

At very different places in the cutting edge of earthquake-proofing technology, rural Pakistan and urban New Zealand (and similarly placed locations elsewhere in the modern world) now have to revise and revisit the planning of construction in terms of settlement densities, architectural construction and mobility patterns if the patterns of past failure are not to be repeated.

Preparing the public, and particularly children, for coping with the circumstances of disasters has begun to be a part of the planning process. Christchurch already had procedures in place when the earthquake struck. Ensuring that children know what to do to survive and reducing trauma for children when routine circumstances are disturbed become important dimensions when routine urban life geographically separates children from their parents and families. When disasters occur, routine urban separation takes a critical form and the potential for traumatic separation is present. Children's crisis mobility arrangements require discussion under routine circumstances and before crises occurs. In Illinois, emergency preparedness for children has taken the form of an electronic game entitled 'The Day the Earth Shook' – within such an arrangement the potential for comprehensively preparing children for crisis mobility exists. Similarly, the application of cell-phone based household based tracking technologies for locating members in crisis given the recent pattern of disasters requires consideration within this framework.

The planning of construction requires the planning of communication to meet disaster conditions: developing robust and flexible communication systems that permit the reconnection in complex urban and semi-urban environments of those endangered, dislocated and separated by disaster becomes an increasingly evident imperative and this is nowhere more apparent than in Japan which was still struggling to locate a missing 16,000 population more than three weeks after its cumulative set of disasters.

## The missing ancestors: Japan and cumulative catastrophe

> RIKUZENTAKATA, Japan -- The elderly couple fled their home on foot as the warning sirens blared. But they could not keep up with their neighbors and fell behind as the tsunami rushed in. Nearly a week later, 71-year-old Taeko Kanno and her husband are still missing. "I think there is no hope," said Katsuo Maiya, Kanno's brother-in-law. "I can't find them. The only thing I can do is wait until the military collects their bodies." As retrieving bodies increasingly becomes the focus of rescue crews in Japan's northeast, it's clear that Friday's earthquake and tsunami – believed to have killed more than 10,000 – took their heaviest toll on the elderly in this rapidly aging nation, where nearly one in four people is over 65. Japan (The Huffington Post 2011).

The quadruple set of disasters occurring in Japan in 2011 (earthquake, tsunami, nuclear meltdown and unseasonal extreme weather) have had dramatic effects. Both the confirmed levels of death (in excess of 10,000) and the levels of persons who remain missing (around 16,000) are at levels that challenge the validity and viability of past planning frameworks for the coastal settlements of Japan. What is evident from the press reportage is that the old fared badly in the race for survival[5] (Booth and McCurry 2011).

> Japan's older population is not evenly distributed across the country. Over the last decades, much of Japan's youth has slowly left the countryside for jobs in the country's main cities. As a result, agricultural regions, such as those affected by the earthquake and tsunami, include some of Japan's oldest populations. Adding to this demographic challenge, media reports are quickly making it apparent that many older Japanese did not have the necessary help to survive the earthquake and the tsunami that followed.

The ancestors are now missing in a society which revered its elders: the whereabouts of their bodies is not even known. Planning in the future has to give more thought to the location of the residences of the elderly and other categories of the vulnerable and to the evacuation and mobility strategies necessary to ensure their survival alongside their younger counterparts.

Alongside the tragedy of the loss of the old, the level of settlement and infrastructure destruction has destroyed the material history of place and survivors are currently scavenging their old settlements for the material culture of their memories without certainty as to even the placement of neighbourhoods.

> "My daughter is missing. She is gone. I was up in the hills above town and couldn't get to her in time," Kikuchi said in a steady voice.

---

5 See http://iahsa.wordpress.com/2011/03/15/japan-earthquake-and-tsunami-likely-to-have-overwhelmingly-affected-older-adults/ [accessed: March 2011].

> "I am looking for something that can remind me of her. There must be something here," Kikuchi said and then resumed wandering through the mud-strewn debris in the town of broken buildings erased by a giant wall of water.
>
> The tsunami took just about everything from those in its path who escaped with just the clothes on their backs, as the wall of water destroyed homes, pictures, keepsakes and all the objects that are touchstones of memories.
>
> People who have spent decades in these northeastern coastal towns now find themselves disoriented when they return because streets have vanished and landmarks disappeared.[6]

The use of new information technologies to reconnect this disorientated population has been less than adequate. Reconnecting cellphone and smart phone technologies can be seen as critical to the reconnecting of the displaced and the ready identification of the missing. However, it does not appear to have been prioritized within the emergency framework. At the international level, data packaging and management systems normally routed through Japan were rerouted through other destinations very rapidly after the emergency so as to not interrupt international commerce. Virtual mobility options require explicit consideration as part of emergency disaster management. Yet the images emerging from Japan indicate the low technology pattern of current strategies of social reconnection: one example includes lists of names pinned up on walls, which formed the basis of social reconnection in a society known for its development of cutting edge communication technologies.

New information technologies will also be important in restoring the memories of those places that have been destroyed. The virtual recreation of neighbourhoods – making use of the memories of survivors and re-integrating the current surviving material fragments into a more closely representative reflection of the past that has been lost – may be an important step in the overcoming of the trauma of 2011. Virtual Japan may be an important mechanism for re-establishing the material Japan that was destroyed by the tsunami.

In redeveloping material Japan along the coast impacted by the tsunami there are important infrastructure lessons to be learnt. Narrow streets provide the tsunami with additional force in its 'glacial' movement of cumulative urban infrastructural debris across the coastal strip and flood plain: Japan's historic preference for narrow streets may need rethinking. Similarly, the earthquake that preceded and triggered this tsunami resulted in a lowering of the land mass so that the height of the sea wall defences against tsunamis was effectively lowered itself; allowing a tsunami wave to cross to land which the defences – had the land not lowered – would have been sufficient to stop. As with the experience in New Zealand, the engineering knowledge in respect of the tsunami sea wall defences failed because it had not envisaged that the land could be lowered in synchronicity with the

6 See http://news.yahoo.com/s/nm/20110322/lf_nm_life/us_quake_japan_victims [accessed March 2011].

advent of the tsunami. The level of the infrastructural debris that accompanied the tsunami and the damage that the cumulative 'glacial' pressure generated raises the issue of the scale and organization of contemporary infrastructure. The prospect of lines of defence rather than the one sea wall is one direction that engineering solutions might explore; the relocation of populations or the redesign of settlement architectures are others.

The development of portable emergency infrastructural technology is another avenue of innovation that is worthy of exploration. Emergency water treatment facilities, emergency power supplies, emergency telecommunication facilities and emergency medical facilities all require thinking about. These are rarely necessary under routine conditions, but are vital in the context of complex disasters. Further, they are not beyond current technological competences and are simply a matter of developing an excess capacity that can be harnessed when necessary. Correspondingly, less centralized and more distributed development of infrastructure would afford higher levels of protection of essential services especially to the most vulnerable.

In ending this discussion of the all too evident challenge presented to sustainable urban and peri-urban mobility and sustainable urban transport by recent events, it is time to acknowledge the 'elephant in the room'. The location of atomic power plants in seismically challenged coastal areas requires major rethinking. The failure of the back-up systems and the absence of deeper back-up systems to compensate for the back-up systems that failed has now left a situation where substantial radiation risks exist not only in the immediate locality, but through the patterns of water connection to a major world city. The metropolis, Tokyo, survived the earthquake, but its interconnectedness through the water infrastructure to the impacted area puts it at continued risk. The consequences of scale are all too evident: large cities are serviced by large facilities and these carry large risks in their potential impacts ranging from unstoppable debris on a tsunami wave to an unstoppable or not quickly stoppable radiation leak.

These are contemporary reflections on tragedy and disaster but they come with contemporary reflections on scale and reorganization, on architectural design and on the role of technology in offering the benefits previously confined to scale and now available on a more localizable and distributed platform. The well-known traditional Japanese tsunami image of the power of the wave, like Dr Foster, provided its own traditional alert: the imperative now may be to harness it.

## Conclusion

Sustainable urban mobility and sustainable urban transport in order to be useful as guiding concepts have to be placed within a framework that can plan routine arrangements so as to provide the maximum flexibility for coping with predictable though not necessarily dateable disasters. There is little evidence, either as planners or as a critical community reflecting on the nature and quality of provision, that we have reached this stage.

From Flood Street New Orleans to the coast of Japan, from Kashmir to Christchurch, from Gloucester into history, the technical tools available to us allow us to chart and to assist in the relaying of relevant knowledge, the sharing of experiences and the search for new patterns of service and settlement that place us all in less danger. For mobility to be sustainable, man's journey has to be made safe. Discussions of social exclusion and social inclusion in respect of transport have to stretch beyond the routine and engage with the prospects of predictable though not necessarily dateable disaster. Inside this discussion the im/mobilities of infrastructure and place itself require new levels of attention.

## References

Aga Khan Development Network 2006 and 2008. Reports [The following reports from Aga Khan Development Network and its global affiliates were chosen to appear in REAL RISK (2006) and RISK WISE (2008); Both titles are books formally published by Tudor Rose for the International Disaster and Risk Conferences (IDRC) with the full support of United Nations]. Available at: http://www.akdn.org/publications/kashmir_earthquake.pdf [accessed: 25 October 2012] and http://www.ice.org.uk/patoolkit/Policy/Sustainable-infrastructure#card_8 CE [accessed: 18 April 2012].

BBC Gloucestershire 2009a. Floods 2007: Remembering the crisis [Online: BBC Gloucestershire]. Available at: http://www.bbc.co.uk/gloucestershire/floods/ [accessed: 25 October 2012].

BBC Gloucestershire 2009b. Floods. The day it rained forever [Online: BBC Gloucestershire]. Available at: http://www.bbc.co.uk/gloucestershire/content/articles/2007/12/28/flood_documentary_feature.shtml [accessed: 25 October 2012].

Booth, R. and McCurry, J. 2011. Japanese earthquake takes heavy toll on ageing population. Shocking stories of deaths emerge as the military is enlisted to help at the Fukushima Daiichi nuclear plant. *The Guardian* [Online: 17 March 2011]. Available at: http://www.guardian.co.uk/world/2011/mar/17/japanese-earthquake-toll-ageing-population-deaths [accessed: 25 October 2012].

De Graaf, R.E. 2009. *Innovations in urban water management to reduce the vulnerability of cities: feasibility, case studies and governance*. Doctoral dissertation. Technical University of Delft: The Netherlands. Available at: http://www.ecoboot.nl/ecoboot_new/?p=351 [accessed: 25 October 2012].

Dibben, K. 2011. Two brothers in ute and tinnie saved 16 from Grantham flood hell. *The Sunday Mail* [Online, 16 January 2011]. Available at: http://www.couriermail.com.au/news/sunday-mail/two-brothers-in-ute-and-tinnie-saved-16-from-grantham-flood-hell/story-e6frep2f-1225988500287 [accessed: 25 October 2012].

Façade 2011. Living on Stilts [Online, 5 January 2011]. Available at: http://thisfacade.com/architecture/living-on-stilts.html [accessed: 25 October 2012].

Freeman, P.K. 2000. Infrastructure, Natural Disasters, and Poverty, in *Managing Disaster Risk in Emerging Economies*, edited by E. Kreimer and M. Arnold. Washington, DC: World Bank, 55–61.

Gloucestershire 2009. We are still at risk from flooding [Online]. Available at: http://www.thisisgloucestershire.co.uk/news/risk-flooding/article-1178143-detail/article.html [accessed: 25 October 2012].

Grieco, M. 2007. Transport and social exclusion: new policy grounds, new policy options, in *Moving Through Nets: The Physical and Social Dimensions of Travel*, edited by K. Axhausen. Amsterdam: Elsevier, 9–32.

Grieco, M. and Hine, J. 2008. Stranded mobilities, human disasters: the interaction of mobility and social exclusion in crisis circumstances, in *The Ethics of Mobilities: Rethinking Place, Exclusion, Freedom and Environment*, edited by S. Bergmann and T. Sager. Farnham: Ashgate Publishing.

Hub Pages 2012. The Demise of the Queenslander – an Ideal House for Floodplains [Online]. Available at: http://hubpages.com/hub/The-Demise-of-the-Queenslander-an-Ideal-House-for-Floodplains [accessed: 25 October 2012].

The Huffington Post 2011. Earthquake 2011: Elderly Hard-Hit As Hope For Missing Fades. *The Huffington Post* [Online, 17 March 2011]. Available at: http://www.huffingtonpost.com/social/moby49/japan-earthquake-2011-elderly-hard-hit_n_837117_81074075.html [accessed: 25 October 2012].

King, A. 1997. Earthquake Loads and Earthquake Resistant Design of Buildings [Online: Building Research Association of New Zealand (BRANZ)]. Available at: http://www.branz.co.nz/cms_show_download.php?id=405a730d9f218fe0b18d4296e28ff5cf7fad52ab [accessed: 25 October 2012].

Litman, T. 2006. Lessons from Katrina and Rita: What Major Disasters Can Teach Transportation Planners. *Journal of Transportation Engineering*, 132, 11–8.

McSaveney, E. 2009. Earthquakes – building for earthquake resistance [Online: Te Ara – the Encyclopedia of New Zealand]. Available at: http://www.TeAra.govt.nz/ en/earthquakes/4 [accessed: 25 October 2012].

Naeem, Dr. A., et al. 2005. A Summary Report on Muzaffarabad Earthquake, Pakistan. *Earthquake Engineering Center at the Department of Civil Engineering, N-W.F.P. University of Engineering and Technology, Peshawar, Pakistan* [Online: reliefweb]. Available at: http://reliefweb.int/report/pakistan/pakistan-summary-report-muzaffarabad-earthquake [accessed: 25 October 2012].

Opie, I. and Opie, P. 1997. The Oxford Dictionary of Nursery Rhymes. 'Doctor Forster'. The Roud Folk Song Index Number of 19288 (Oxford: Oxford University Press, 1951, 2nd edn., 1997), 173.

Padgett, T. 2010. New Orleans' Lower Ninth: Katrina's Forgotten Victim? *Time* [Online, 27 August 2010]. Available at: http://www.time.com/time/specials/packages/article/0,28804,2012217_2012252_2012673,00.html [accessed: 25 October 2012].

# Chapter 13
# MaxSumo and MaxEva: Tools for Measuring the Success of Mobility Management Projects

Fred Dotter

## Introduction

### *A definition of Mobility Management*

The definition of Mobility Management and the categorization of Mobility Management measures as approved by the MAX-consortium[1] and EPOMM[2] are as follows:

> Mobility Management (MM) is a concept to promote sustainable transport and manage the demand for car use by changing travellers' attitudes and behaviour. At the core of Mobility Management are 'soft' measures like information and communication, organising services and coordinating activities of different partners. 'Soft' measures most often enhance the effectiveness of 'hard' measures within urban transport (e.g., new tram lines, new roads and new bike lanes). Mobility Management measures (in comparison to 'hard', technological, measures) do not necessarily require large financial investments and may have a high benefit-cost ratio (Baasch et al. 2007).

The objective of Mobility Management (MM) is to induce citizens to practice 'reasoned mobility', that is, organizing their daily activity patterns in an efficient way, in terms of costs, energy savings and environment conservation.

In many European countries Mobility Management is becoming an accepted approach for addressing different kinds of transport related questions such as congestion, modal shift and emissions. This means that the demand for structured ways of identifying, implementing and evaluating different kinds of Mobility

1 MAX Successful Travel Awareness Campaigns and Mobility Management Strategies (www.max-success.eu) ran from 2006 to 2009 and was the largest and one of the most important multinational research projects on Mobility Management and Travel Awareness in transport, funded by the EU 6th Framework Programme.

2 EPOMM – European Platform on Mobility Management, www.epomm.eu.

Management measures is increasing. Furthermore, the use of common methods and routines will make comparisons between countries easier as well as making the introduction of Mobility Management easier in countries where the potential of Mobility Management is not widely known.

*Understanding mobility behaviour change*

The primary purpose of any evaluation is to demonstrate that the mobility project implemented has achieved its main aims. In the case of Mobility Management this refers to whether the project has succeeded in making individuals' travel behaviour more sustainable.

In order to successfully change people's behaviour, it is essential to understand the underlying processes necessary for behavioural change to occur, and to use this knowledge for implementation strategies. Two key facts related to behavioural change are most relevant to the implementation and evaluation of mobility projects:

- First, in any given population some people are more susceptible to changing their travel behaviour than others. This partly relates to more subjective factors such as their attitudes and perceptions towards their current travel choices. For other people the barriers to modal shift are more objective, such as if there is no bus service operating on the route for their journey, or if they have a disability that prevents them switching car trips to cycling or walking.
- Secondly, it is increasingly acknowledged that in many instances behavioural change does not occur as a one-step process. Instead it can be viewed as a series of stages (or steps) that individuals progress through in order to reach the final stage: a new habitual behaviour. Thus, more subtle changes in attitudes and perceptions towards alternative modes (reflecting a greater propensity to change behaviour) will occur simultaneously to obvious behavioural changes.

In order to obtain a 'fuller picture' of what the intervention has achieved it is important to measure more subtle changes in attitudes and perceptions as well as overt behavioural change *per se*. Evaluations that focus solely on actual behavioural change will not show this and therefore their success in moving people towards a behavioural change would be understated. Measurements of people's 'stage position' can be used before the project starts to help inform the selection and design of subsequent measures that may provide the 'final push' resulting in the ultimate goal of actual behavioural change. For example, a mass-media travel awareness initiative may increase some individual awareness of and propensity to use alternative transport modes. The information provided may inspire individuals to consider the possibility of using these alternative modes, although further initiatives (perhaps personalized travel advice or reduced fares) may be necessary for them to actually switch to other modes.

*Standardization of evaluation tools*

For anyone carrying out Mobility Management projects, it is of primary interest to know and to be able to show that the effort and the costs invested are justified. Still, many projects are not evaluated at all. One obvious reason is the lack of a common and generally accepted evaluation tool.

A number of ex-post assessment tools for Travel Awareness and Mobility Management related activities have been developed in recent years (MAX 2007), building on the research work done in EU projects.

The most interesting examples are:

- The Monitoring and Evaluation Toolkit of the MOST project (MOST-MET);
- The Swedish System for Evaluation of Mobility Projects (SUMO[3]) and the database SARA;
- The Campaign Assessment Guidance (CAG), of the TAPESTRY project.

These tools are the result of previous European research projects and are used to measure different indicators and stages of change and to compare the situation before and after a MM measure is implemented. Further, the knowledge the results contain may assist in the formation of a prospective assessment tool (Hyllenius et al. 2009).

There are strong similarities among the assessment levels used in MOST-MET, SUMO and CAG. As they are all meant to demonstrate the individual step-by-step development of the target group members (from awareness raising, thinking about the issue, accepting the alternatives offered and finally to a change in behaviour), their similarities are not at all surprising. Nevertheless, there are some important differences. Whereas MOST-MET and SUMO start with awareness of the offered Mobility Management services, CAG concentrates on the awareness of the problems addressed in the campaigns and the acceptance of some personal responsibility for the consequences arising from individual behaviour. Unlike MOST-MET and SUMO though, CAG – as a campaign assessment tool – is not able to measure the demand for and satisfaction with Mobility Management services as no services are offered in a campaign. Therefore, a tool made for both Mobility Management and campaigns should include all measures, as in the case of MaxSumo.

## MaxSumo and MaxEva

The rationale behind the development of MaxSumo was to make the effects of MM measures more predictable. A state-of-the-art analysis carried out by the

3 SUMO – System for Evaluation of Mobility Projects (for more information, see Hyllenius et al. 2004).

MAX project found that due to the lack of high-quality data available on the effectiveness of MM measures, evaluation studies of a high standard are needed. MAX therefore seeks to provide a structure for applying evaluation methods. It includes randomized control groups sufficient enough to satisfy the most methodologically rigorous psychologists and transport researchers and, more importantly, a model for obtaining reliable evaluation results.

MaxSumo is meant to be a tool by which baseline data on the impact of MM measures and programmes in various fields (for example, worksite-based, school-based, regional programmes or mobility centres) can be gathered on a large scale. This constitutes a central pre-condition for the development and implementation of a prospective assessment tool for Mobility Management programmes.

Thus, MaxSumo seeks to set up a framework for a prospective assessment tool that can then be developed in the future when sufficient baseline data is available.

By developing MaxSumo a new, enhanced tool for planning, monitoring and evaluating mobility projects now exists. Some central improvements that could be achieved with MaxSumo are the inclusion of:

- Recent research results about the behavioural change process with regard to modal choice and modal shift.
- Brief information about designing an evaluation and establishing cause-and-effect, in order to raise awareness about the importance of evaluation *per se* and of the different evaluation designs.
- Concrete examples about data collection, suitable questions and other methods (for example a mobility-diary). These provide guidance for monitoring and evaluation and are expected to improve the quality of evaluations as well as the ability to compare different evaluations.

The online database MaxEva,[4] particularly designed to support users of MaxSumo, stores monitoring and evaluation data and helps users especially (in the future) to benchmark their own project results with others. It uses the type of data generated from a MaxSumo structured evaluation. As a web version of MaxSumo, it provides an easy overview of both interim and final results.

MaxEva simplifies the work by swiftly leading the user to the essential questions. It also provides access to many other examples so that users can see how others have proceeded through the various steps of MaxSumo. The use of MaxEva helps to identify key performance indicators that can help decision-makers to prioritize investments in MM.

MaxEva automatically calculates $CO_2$ emissions by using several selectable, country-specific default values on average $CO_2$ emissions of different transport modes (MAX 2010). This web database contains a list of conversion factors that are multiplied with the change in kilometres per year for each mode. This

4 For more information on MaxEva, please visit www.epomm.eu/maxeva.

provides both the overall results and the result per mode in tonnes of $CO_2$ reduced or increased per year.[5] In MaxEva, conversion factors for countries and years can also be selected. The output of MaxEva includes a results page for each project presenting a quick overview as well as more detailed results (accessible with one or two mouse clicks).[6] The users of MaxEva can approve their own project data at the final stage and allow others to take a look at these projects results through the extensive benchmarking functions of the database as well as through a comfortable map function that localizes all projects.

The final purpose of MaxEva is to accumulate data from a large number of MM projects in order to develop a database with reliable information. Users are then able to use the database to help determine what effects can be expected from MM measures they are considering.

## The MaxSumo process

MaxSumo is a resource providing advice on issues that need to be monitored and evaluated as well as their levels along with giving guidance throughout the entire project, from deciding on the initial necessary steps to finalizing the evaluation. It can thus be seen as a process to help manage project issues and to attain real results. Part of this process deals with aspects that arise at the start of the project, while other processes focus on the running project or the final evaluation stage. Therefore, possible deviations can be corrected at an early stage.

It is important to follow the process intended for the start of the project (to be discussed on the following pages) and to work with the assessment levels at the *outset* of a project in order to be able to carry out monitoring and evaluation at these levels. Following this process also helps to steer the project in a systematic and methodical way towards the correct target group, organization and services (the methods necessary to induce an impact). The use of the different MaxSumo assessment levels must therefore be planned from the very beginning of the project. During the project, feedback from the evaluation provides opportunities to make improvements to the project in progress. This is a very important part of the process (marked with feedback arrows to the right in Figure 13.1). Cause and effect relationships are also observed at the evaluation stage.

---

5 For many countries, these values are not available; in that case MaxEva suggests a European average default value. MaxEva Emission factors: grams $CO_2$ per passenger kilometre (g/pkm) (MAX 2010).

6 MAX Successful Travel Awareness Campaigns and Mobility Management Strategies. Mobility Management at the Austrian Ministry of Environment. Available at: http://www.epomm.eu/maxeva/index.php?id=2&proj_id=25 [accessed: 22 November 2012].

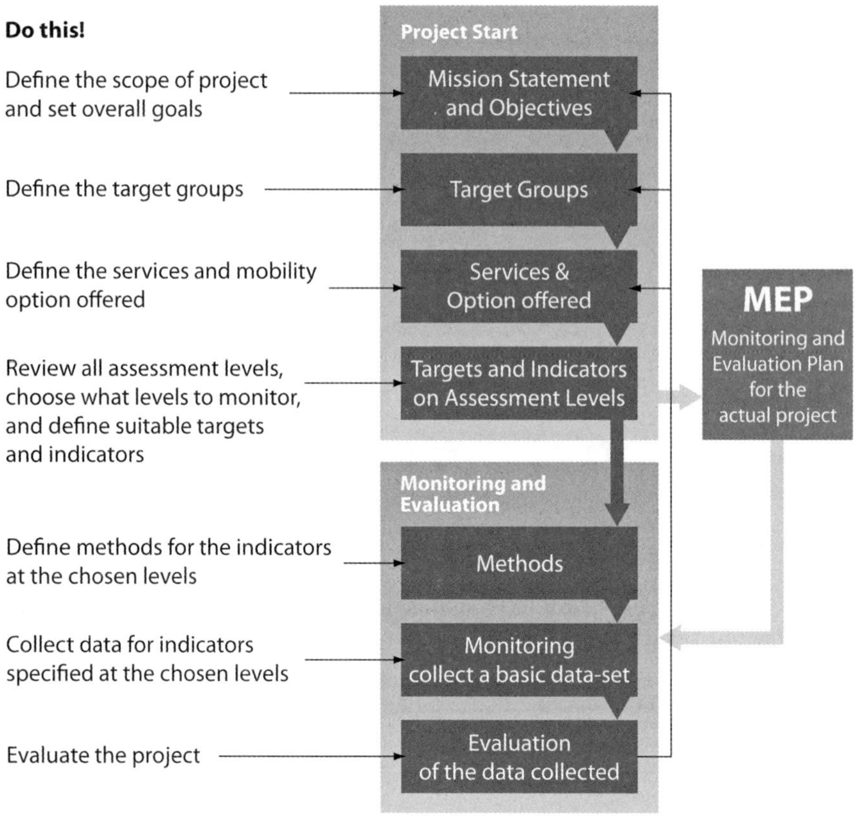


**Figure 13.1 The MaxSumo process: a multi-step procedure initiated when starting to plan a project**

*Source:* MaxSumo: Guidance on how to plan, monitor and evaluate mobility projects (Hyllenius et al. 2009).

To illustrate how MaxSumo is used in reality during the entire evaluation process, a description of a project – the 'test traveller project' – is provided above. The following steps are described and illustrated with the test traveller project in order to better comprehend the theoretical outline.

*The test traveller project ran from September 2011 to October 2011 and was implemented for Hallandstrafiken in Sweden with an overall budget of EUR 10,000.00. Hallandstrafiken AB is the organization responsible for public transport in the county of Halland. It offers locals and visitors attractive travelling opportunities at competitive rates with approximately 11 million journeys per year within the county. The project focused on company employees who normally commute to work by car. The test traveller project offered them the opportunity to try commuting by public transport for one month. The car commuters received information about appropriate public transport services together with free bus tickets, valid for the entire 'test-travel' period. In return for the ticket, participants had to agree to attend meetings, participate in before and after surveys and to actually try out public transport alternatives by using the tickets provided.*

*Step 1: Define the scope and set the overall goals of the project*

In order to successfully monitor and evaluate projects (in order to see if they have reached their objective), it is important to clarify what the project is attempting to do (that is, its target goals) before the project actually begins. These targets must be set in an appropriate and constructive way.

*Test traveller project:*

*The project involved persuading car commuters to break their habit and to commute to work by bus in order to reduce the environmental impact of car travel and congestion on city roads. The overall goal for local authorities was to achieve a more sustainable environment and to reduce road congestion. The overall goal of the transport provider was to be able to entice more people to commute by bus. They were interested in finding out the opinions of people who were not used to taking the bus in order to be able to improve commuting facilities and to ultimately increase the number of bus commuters.*

*Step 2: Define the target groups*

It is important to decide upon and to define a specific target group whose behaviour the project is attempting to change, which in part allows the researcher to focus on using given resources for the groups most likely to achieve the greatest change. The target group will be largely dependent on the aims of the mobility projects. Sometimes there are direct and indirect target groups, such as corporate management or union representatives (indirect target groups) who need to accept the measure in order to reach the direct target group (for example, employees and management staff who are asked to change their behaviour). Often, it is crucial for the success of the implemented measures that the indirect target group is identified and directly involved. Services in the project can be aimed at both groups, but the mobility option offered is only for the direct target group.

*Test traveller project:*

*The project focused on car commuters from the cities of Getinge, Gullbrandstorp, Haverdal and Falkenberg, all small towns of less than 50,000 inhabitants. Out of these, the project particularly targeted 41 adult employees of both genders living in areas with good bus connections between their homes and workplaces, even though the employees currently travel to work by car. In the test traveller project all participants were positive about trying public transport for the trial period.*

*Step 3: Define the mobility services provided and mobility options offered*

Mobility projects aim to promote a change in mobility behaviour by providing mobility services that motivate people to accept and use a new mobility option offered to them. Mobility projects often consist of a range of measures which can be categorized into different mobility services provided by the project or the programme itself. The mobility options offered to the participants support the kind of travel behaviour the project promotes, such as a shift from car use to walking, cycling or public transport; a shift from owning a private car to utilizing car-sharing options; or a reduction by commuters to only use their car four days out of the week instead of five.

*Test traveller project:*

*The mobility service provided was a combination of different kinds of information and a public transport ticket for a trial period. Meetings were held at the companies these commuters work at to recruit car commuters for the trial period. The project provided detailed advice, free public transport passes, and so on, for car commuters that signed up to test public transport. The mobility option offered was intended to encourage people to commute by public transport rather than by car.*

*Step 4: Set targets and define indicators for MaxSumo assessment levels*

Often, many steps are required in a MM project to finally arrive at the desired sustainable transport behaviour. Directly measuring overall goals is often difficult or impossible and furthermore takes a long time. Therefore, using MaxSumo to make measurements at a number of early stages in the project can provide valuable information as the project progresses. When setting targets and deciding on indicators, it is advisable to decide on assessment level(s) that will be included in the project monitoring and evaluation.

*Test traveller project:*

*The project's target (and only indicator) was to achieve a reduction of at least 20 tons of carbon dioxide after one year of being introduced in the cities of Getinge, Gullbrandstorp, Haverdal and Falkenberg. Since evaluating the change in $CO_2$ emissions resulting from the project was the key objective of the monitoring of the project, the most relevant level to focus on was travel behaviour (see below 'Level H: Long-term attitudes and behaviour').*

*Step 5: Define data collection methods*

Once the targets and indicators to be monitored are defined, an appropriate method for collecting the data must also be decided. Many different types of data collection methods exist, including surveys, interviews, site investigations, documentation and counts. The data collection method choice depends on the type of data relevant to the project. During the planning stage, how often data is collected should also be defined.

*Test traveller project:*

*The method used in the test traveller project was a combination of documenting project outputs and activities and using information obtained from questionnaires. Data from before the intervention was collected for modal split and travel distance per mode of the target group (one way trips). This collection was carried out by a questionnaire which achieved a 100 per cent response rate. The summary of the data collection can be found below in Step 7, which provides an overview on the evaluation of collected data.*

*Step 6: Monitor the chosen assessment levels*

When monitoring the project, data collection is recorded based on to the MEP (Monitoring and Evaluation Plan) for the actual project. The MEP shows which indicators need to be measured, which methods need to be used and when measurements need to be taken. The assessment level 'services provided by the project' (see 'Levels A–D' below) should be monitored during the course of the project. Levels A–D all assess how well the services of your project are known or accepted. They are used to measure the project's progress and to judge whether it is on track or if it needs to be redirected to reach the overall project targets.

Background data and current travel behaviour normally need to be collected either before or at the very beginning of the project. If, on the other hand, data is only collected at the end of the project and for the target group alone (without the use of a control group), proper evaluation of intervention effects will be nearly impossible. Travel behaviour after the intervention should always be monitored. If there are behavioural changes in the intervention group, these need to be explained and data on specific background conditions is crucial for doing so. In order to estimate the cost efficiency of various measures as well as the cost effectiveness of the overall project, it is also very helpful to document and to monitor the costs of different measures and services.

*Test traveller project:*

*Here data collected during the intervention regarding modal split and travel distance per mode in the target group was used and compared to data collected before the intervention. The response rate for this was 68 per cent. A summary of the data collected can be found below in Step 7, which provides an overview of evaluating collected data.*

*Step 7: Evaluate the project and explain the observed changes*

When behaviour with or without the project intervention is to be monitored and evaluated, it is important to carefully consider what needs to be compared. If behaviour change is measured in a population or target group before, during and after an intervention or awareness campaign, this population group needs to be compared. Behavioural data from a comparable group should be used if the monitored population has changed during the intervention. Otherwise, monitored effects might be attributed to the intervention when they are actually due to the fact that different populations have been compared. If the groups differ in composition or other behavioural conditions then the difference in mobility behaviour might not be fully attributable to the intervention; it is also important to consider that it could also be caused by other differences.

An evaluation of the behavioural change effects or an early assessment of the project services provided can be performed during or after completion of the project or intervention. In either case the evaluation aims to explain why changes have occurred. To be able to do this care must be taken when examining the possible causes of change. Changes in mobility behaviour could result from factors such as weather, welfare, income and changes in petrol prices, supply of public transport or similar. One of the most important long-term objectives for evaluation is to acquire basic data on cause and effect relationships of the implemented measures. By using MaxSumo, the effects of specific measures at each level can thus be estimated and used to produce a system including all the experiences previously collected.

*Test traveller project:*

*First of all, when it came to travel behaviour, data in MaxEva was used to automatically calculate the resulting effects of the test traveller project, as related to Steps 5 and 6. Data on modal split and travel distance per mode of the target group after the intervention were collected and compared to the data collected before the intervention. This was done using a questionnaire with a 73 per cent response rate and showed in a shift from car use of –27.7 per cent.*

**Table 13.1 Modal share before (Step 5), during (Step 6) and after (Step 7) the intervention in the test traveller project regarding the mode of transport and the length of commuter trips to and from the workplace per day**

| Mode of transport | Modal share before intervention (step 5) | | Modal share during intervention (step 6) | | Modal share after intervention (step 7) | |
|---|---|---|---|---|---|---|
| | Share in % | Km | Share in % | Km | Share in % | Km |
| Car Driver | 90.5 | 17 | ↘ 39.9 | 17 | ↘ 62.8 | 17 |
| Car Passenger | – | – | – | – | – | – |
| Walking | – | – | – | – | – | – |
| Cycling | 5.6 | 7 | ↘ 3.6 | 7 | ↗ 19.4 | 8 |
| Bus | 2.5 | 10 | ↗ 46.4 | 10 | ↗ 14.7 | 13 |
| Train/Tram/Metro | 1.4 | 35 | ↗ 10.1 | 35 | ↗ 3.1 | 41 |
| Motorcycle/Scooter | – | – | – | – | – | – |
| Zero-emissions vehicle | – | – | – | – | – | – |
| | 100% | | 100% | | 100% | |

*In addition to this, the effects of $CO_2$ emission were calculated and the total amount of reduced $CO_2$ emissions was found. The reduction of 10.60 tons of $CO_2$ per year is based on the calculation which was carried out automatically by MaxEva and based on data provided on changes in travel behaviour, the size of the target group and the selected country specific emission factors given for a specific year.*

**Table 13.2 Overview of the available emission factors for Sweden (2010)**

| Mode | Factor | Emission factors based on |
|---|---|---|
| Car Driver | 183 | Average models: 221 g/km well to wheel, 1.2 people/car[11] |
| Car Passenger | 100 | Average models: 221 g/km well to wheel, 1.2 people/car |
| Walking | 0 | – |
| Cycling | 0 | – |
| Bus | 83 | FRIDA: Average buses: 974 g/km, SIKA: 11.73 people/bus[12] |
| Train/Tram/Metro | 5 | Most trains in Sweden are run with renewable electricity[13] |
| Zero-emissions vehicle | 0 | – |
| Motorcycle/Scooter | 112 | MC and moped, average[14] |

*Source:* MAX 2010

**Table 13.3 Mileage effect in kilometres (km)**

| Mode | Before | During | After |
|---|---|---|---|
| Car Driver | 227083 km | 100117 km (–126966 km) | 157578 km (–69505 km) |
| Car Passenger | 0 | 0 | 0 |
| Walking | 0 | 0 | 0 |
| Cycling | 57856 km | 3720 km (–2066 km) | 22908 km (+17122 km) |
| Bus | 3690 km | 89032 km (+85342 km) | 28206 km (+24516 km) |
| Train/Tram/Metro | 7232 km | 62612 km (+55380 km) | 18760 km (+11528 km) |
| Motorcycle/ Scooter | 0 | 0 | 0 |
| Zero-emissions vehicle | 0 | 0 | 0 |

## The MaxSumo assessment levels

The 'gap' between what is achieved in a mobility project and the desired target for system impact is often huge. The figure below shows that MaxSumo divides this gap into smaller steps ('assessment levels') that can actually be monitored and evaluated. It provides step-by-step guidance from the start of the project through the implementation and all the way to system impacts. Each step moves the evaluation closer to the overall target and shows the impacts and effects on a systematic level.

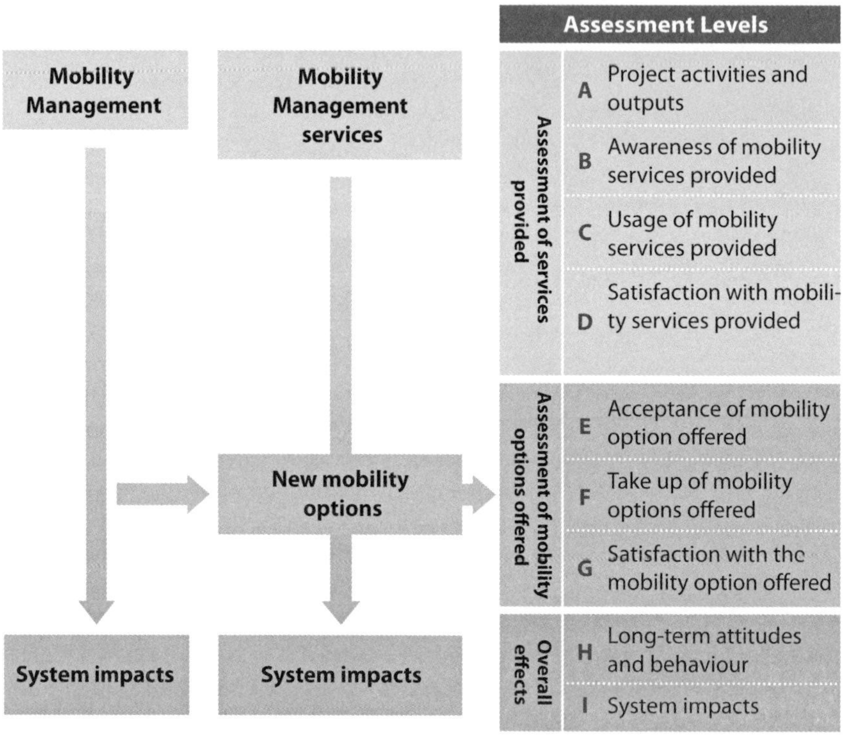


**Figure 13.2 MaxSumo divides the gap between what we do in MM and the targeted impacts into manageable steps ('assessment levels')**

*Source:* MaxSumo: Guidance on how to plan, monitor and evaluate mobility projects (Hyllenius et al. 2009).

The different MaxSumo levels follow each other in a logical way. At each level a decision needs to be made on what is measured and how, which indicators are to be used and how they will be measured. The assessment levels are co-ordinated in such a systematic way as to allow the direct monitoring of the effects of the project during implementation.

The different assessment levels in MaxSumo are divided into four main categories:

- 'Intervention framework conditions', which refers to the underlying background conditions at the project location and the group characteristics the project focuses on.
- 'Services provided by the project', which refers to the different activities and outputs the project utilizes in order to promote changes in travel behaviour.
- 'Mobility options offered through the services provided', which refers to the kind of travel behaviour the project is attempting to persuade the target sample to use (for example, the attempt to get motorists to switch to public transport, cycling or walking for some or all of their trips).
- 'Overall effects', which refers to the main outcomes of the project with regard to new attitudes and behaviour adopted, as well as system impacts achieved by a change in behaviour (for example, reduction in car mileage, $CO_2$ emissions).

*Level 0: External factors and person-related factors*

External factors are the local conditions that apply to the people targeted by the mobility project measures. These factors can vary, for example, due to the current political situation or legislation (for example, a local authority or government may introduce new laws), the current price of fuel and public transport (which could increase or decrease), and so on. In such instances these external factors are likely to influence modal choice and, unless they are documented and ideally controlled, it will be hard to accurately establish the precise impacts of the mobility project (the subject of the evaluation). Most of the monitoring data for these indicators can be gathered via official sources such as official statistics or online reports. Alternatively, they can be obtained via secondary data sources, such as interviews with key personnel at local authorities or businesses, or during the preliminary assessment performed as part of the mobility project.

Person-related factors refer to the personal characteristics of the people targeted by the mobility project. They can be broadly classified as either objective or subjective person-related factors. Objective person-related factors include age, gender and living and working conditions, among others. Subjective person-related factors refer to the individuals' intrinsic values, attitudes or stage position. Most of the data for both objective and subjective person-related factors can be gathered via questionnaire surveys or interviews conducted with people in the target population group.

*Test traveller project:*

*Before the start of the project investigations regarding the location of the company and the accessibility to public transport were carried out. Furthermore, in the course of the survey, person-related data (age, gender) were collected.*

| | | |
|---|---|---|
| **Intervention framework conditions** | | **External factors**<br>A description of the external conditions for the measures. These are the same for all users. |
| | | **Person-related factors**<br>Information about the personal situation for different individuals, in order to divide them into target groups, etc. This includes both „objective conditions" (such as distance home-to-work, access to public transport) and „subjective conditions" (such as the individual's stage of behavioural change and the travel behaviour before the intervention). |
| **Assessment of services provided** | A | **Project activities and outputs**<br>Describes the project effort invested in the measure to change behaviour, such as meetings, material distributed, data systems introduced, decisions on travel policy, etc, and the costs for this. |
| | B | **Awareness of mobility services provided**<br>Describes the awareness of the project or the mobility provided. |
| | C | **Usage of mobility services provided**<br>Among those people that are aware of the services, this level describes the usage or the interest shown in the project or mobility services. |
| | D | **Satisfaction with mobility services provided**<br>Measures how satisfied users are with the services provided. |
| **Assessment of mobility options offered** | E | **Acceptance of mobility option offered**<br>Describes the acceptance of the mobility option offered, by measuring the intent to change behaviour. |
| | F | **Take up of mobility options offered**<br>Measures how many test new mobility options offered, i.e. try a new mobility behaviour. |
| | G | **Satisfaction with the mobility option offered**<br>Shows if people that have tested the mobility option offered are satisfied with it (often a pre-condition if they are to make it a permanent behavioural change). |
| **Overall effects** | H | **Long-term attitudes and behaviour**<br>Measures how many users, due to the intervention, adopt new attitudes and ultimately change how they travel. |
| | I | **System impacts**<br>These are effects that the project or programme is aiming for at the system level, e.g. effect on total traffic on an urban road network. It assesses e.g. how much vehicle mileage, emissions, energy consumption or accidents have changed as a result of the change of travel behaviour. |

**Figure 13.3 MaxSumo assessment levels**

*Source:* MaxSumo: Guidance on how to plan, monitor and evaluate mobility projects.

*Level A: Project activities and outputs*

Project outputs and activities refer to the various activities associated with the implemented mobility project measures and vary from project to project. It is particularly important to monitor all actual activities, especially when it is necessary to overcome barriers and when the project requires a concerted effort and investment during the initial phase (for example, if employees need to be convinced to participate in the project). Outputs and activities are measured by continuously and consistently recording each implementation phase and by quantifying the volume of activities through the use of simple summaries to keep track of how many people have been reached by certain information (documentation) or meetings.

*Test traveller project:*

*Apart from internal information and the survey, no further project activities have been undertaken to promote the project.*

*Level B: Awareness of the mobility services provided*

This level refers to the grade of awareness of the different mobility services provided. It is important to collect information at this level because awareness about the project's existence is, in most cases, a basic prerequisite for anyone who considers using the services the project is promoting. Measuring awareness of the service provides information on how well and widely-known it is as well as how successful the marketing has been. A good level of project awareness is valuable in itself, even if not everyone can or wants to participate in the project. Measuring awareness can also provide early feedback if the project is not achieving its potential impact. If awareness is relatively low, this might highlight the need for further or different types of awareness raising activities in order to make the scheme more successful. Measuring awareness can be done via interviews with people whose behaviour the mobility project is aiming to change.

*Test traveller project:*

*Evaluation showed that 100 per cent of the target group was aware of the services.*

*Level C: Usage of the mobility services provided*

This level refers to the number of people using the services provided within the mobility project. Measuring this number helps to determine how effective the efforts have been. Measuring usage also provides an opportunity to obtain feedback on project activities. Using the provided services is a first step towards a change in behaviour, but it does not guarantee long-term change. Usage of the services provided by a MM-project is measured by counting and documenting how many people have used the service.

*Test traveller project:*

*Evaluation showed that 100 per cent of the target group used the services provided.*

*Level D: Satisfaction with the mobility services provided*

Satisfaction with mobility services provided refers to the level of satisfaction with the mobility service provided and reflects how well the services meet the needs of the target group. This can be measured through follow-up interviews with people who have used the service and provides useful information about how the services can be improved to better meet their needs.

*Test traveller project:*

*Evaluation showed that 97.56 per cent of the target group was satisfied with the services provided.*

*Level E: Acceptance of the mobility options offered*

Acceptance of the mobility option offered expresses a willingness to test it. But acceptance of an option offered does not necessarily mean that people will actually test the option or further change their behaviour to use the offered mobility mode in the future.

*Test traveller project:*

*Evaluation showed that 100 per cent of the target group agreed to try the mobility options offered.*

*Level F: Take-up of the mobility options offered*

This level refers to the number of people who try out or test the mobility option offered. In most mobility projects, individuals are encouraged to test new travel behaviour. In MaxSumo, this is measured as 'take-up of mobility options offered'. Testing and trying a new mode should not be confused with long-term behavioural changes though. If it is not possible to monitor this level during the project period, it could be asked for retrospectively in the after-survey. Thus it would be measured together with longer-term behaviour, which is a necessary part of the project evaluation.

*Test traveller project:*

*Evaluation showed that 90.24 per cent of the target group tried the mobility options offered.*

*Level G: Satisfaction with the mobility options offered*

A certain level of satisfaction with the new type of behaviour the participants have tested is an important prerequisite for recurring and hopefully long-term behavioural change. Measuring the satisfaction with the mobility options offered also provides useful information about how the options could be improved to better meet the needs of the users.

*Test traveller project:*

*Evaluation showed that 70.73 per cent of the target group was satisfied with the options offered.*

*Level H: Long-term attitudes and behaviour*

The typical overall goal of mobility projects is to change individual travel behaviour to more sustainable transport modes on a long-term basis. Therefore, changes in attitudes are of interest to the project. Here attitudes refer to a person's position or intrinsic values towards transport modes. Behaviour, on the other hand, refers to a person's modal choice for all or specific trips they make. Information about behaviour after the completion of a project can be obtained through questionnaire surveys or interviews with the participants. These surveys should take place several months, but preferably one year after the end of the intervention in order to measure long-term changes.

*Test traveller project:*

*Evaluation after one year showed that 26.83 per cent of the target group used the options offered at least once a week for the entire year following the intervention period. Furthermore, a reduction of approximately 9 tonnes of $CO_2$ per year on a permanent basis could be expected and 40 other people were inspired to test public transport.*

*Level I: System impacts*

Long-term changes in individual modal choice can provide lasting impacts on the road transport system, including reduced air pollutants, noise emissions, vehicle mileage and energy consumption, as well as fewer accidents and a reduction in the space taken by parked cars. The sum of these changes within a target group shows the impact at a system level. This is estimated using data collected at previous levels. MaxSumo suggests calculating the impact of the project in terms of emission reductions. It is often difficult to show the effects of a (small) mobility project on the actual air quality when using ambient measurements of pollutants. This is due to the fact that the type or size of change can be relatively small (especially in the short-term) in comparison to the overall emissions situation. In addition, measurement of ambient concentrations only shows the overall levels of pollutants in the air rather than changes in quantities discharged from different sources. In the case of carbon dioxide, only the total emissions per source matter as local concentrations are irrelevant. To facilitate a comparison between projects it may be more useful to focus on one or a few pollutants. $CO_2$ is a key policy challenge, so it should be included in every report written on a sustainable mobility project. Other substances of interest could be nitrous oxides and particulate matter, which are also considered important for local air quality.

If a one-group design is used, system impacts are calculated by comparing the situation before and after a project. When a comparison or control group design is used, system impacts are calculated by comparing the situation of the intervention and control/comparison group after a project.

*Test traveller project:*

*In the test traveller project the one-group design was used. Only one group existed and the system impact of the project referred to the calculated reduction of 10.60* $CO_2$ *per year, based on the calculation carried out automatically by MaxEva (see also MaxSumo process, Step 7).*

## Summary and conclusions

To summarize, MaxSumo aims to increase the number of evaluated Mobility Management projects and to provide guidance on the most important aspects of a 'good' evaluation. It also helps in structuring, planning and implementing a project in such a way that monitoring and evaluation are an integral part of the project. Since MaxSumo is mainly aimed at practitioners (who do not necessarily carry out the evaluation themselves), it is important to keep the text as short, simple and compact as possible. Therefore, information on special tasks such as statistical data analyses are not included.

MaxSumo can be accessed through both the MAX webpage and the EPOMM website. It is therefore expected that MaxSumo will be a recommended tool for the evaluation in future MM projects in different European countries and in MM related EU projects.

MaxSumo covers the most important issues connected to the evaluation of MM projects. However, the importance of trainings and individualized feedback should not be underestimated, as shown through the use of SUMO in Sweden and through earlier testing of MaxSumo. Support for both could possibly be achieved through EPOMM trainings and a helpdesk function or via the EPOMM PLUS project.

With regard to MaxEva, it is important to keep in mind that this database is not an evaluation guide or handbook in itself. It was developed to store evaluation data and was specifically designed to suit users of MaxSumo. The only way this tool will become truly useful in the field of Mobility Management is if it is used by as many users as possible. Unfortunately, the benefits of this tool can only be shown after it has been used. It is expected that the use of MaxEva will help to identify where further developments and improvements are necessary in the future.

As an interactive web database for evaluating data on Mobility Management projects, it is essential that the MaxEva database is maintained and further developed. MaxEva needs consistent maintenance of the algorithms calculating the exhaust emission of carbon dioxide since this calculation is dependent on energy and fuel consumption as well as the mix of fuels consumed by transport (users). Also, carbon dioxide emissions from cars, buses and trains are continuously changing as new cars, fuels and other energy sources are used. As a result, the algorithms of MaxEva need to be updated regularly in order to keep up with such changes.

Questions about the user-friendliness and usability of MaxEva remain and therefore the program needs to be tested so that it can be improved. Incentives need to be provided to MM practitioners so that they begin to use MaxEva. Unfortunately though, at this stage, it is still unclear what these should be. As for MaxSumo, the possible need for trainings and individualized feedback from users of MaxEva should not be underestimated. Experience with applying SARA in Sweden shows a need for both trainings and individual feedback during monitoring and evaluation. This could possibly be solved via combined trainings in MaxSumo and MaxEva as well as a helpdesk function at EPOMM or via the EPOMM PLUS project.

## References


Baasch, S., Bossaert, E., Ljungberg, Ch., Papadimitriou, S., Papaioannou, P., Posch, K.H., Rye, T. and Welsch, J. 2007. *Definition of Mobility Management* [Online: European Platform on Mobility Management (EPOMM)]. Available at: http://www.epomm.eu/index.php?id=2590 [accessed: 22 November 2012].

Hyllenius, P., Ljungberg, C., Smidfelt Rosqvist, L. and Trivector Traffic 2004. *SUMO – Systems for Evaluation of Mobility Projects* [Online: EU Added Value]. Available at: http://eu-added-value.eu/docs/Evaluation_SUMO_en.pdf [accessed: 23 November 2012].

Hyllenius, P., Smidfelt Rosqvist, L., Haustein, S., Welsch, J., Carreno, M. and Rye, T. 2009. *MaxSum – Guidance on how to plan, monitor and evaluate mobility projects*. Brussels: MAX – Successful Travel Awareness Campaigns and Mobility Management Strategies.

MAX 2007. *Comprehensive State of the Art Report, Annex B1.2, Assessment Tools* [Online]. Available at: http://www.max-success.eu/downloads/MAX_SoA_AnnexB1_2.pdf [accessed: 22 November 2012].

MAX 2010. *MaxEva Emission factors: grams CO2 per passenger kilometer (g/pkm)* [Online: European Platform on Mobility Management (EPOMM)]. Available at: http://www.epomm.eu/maxeva/helptext/maxeva_emissions.pdf [accessed: 22 November 2012].


## Projects and websites

EPOMM – European Platform on Mobility Management: http://www.epomm.eu [accessed: 22 November 2012].

MAX Successful Travel Awareness Campaigns and Mobility Management Strategies: http://www.max-success.eu [accessed: 22 November 2012].

MaxEva – the EPOMM evaluation tool: http://www.epomm.eu/maxeva [accessed: 22. November 2012].

# Index

References to illustrations are in **bold**.